Advancing Maths for AQA

Revise for CORE 1

Tony Clough

Series editors
Sam Boardman   **Roger Williamson**   **Ted Graham**

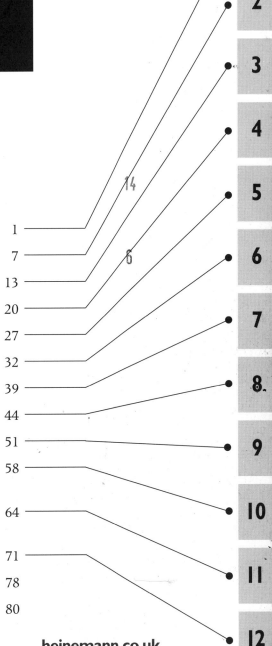

heinemann.co.uk
✓ Free online support
✓ Useful weblinks
✓ 24 hour online ordering

01865 888058

Heinemann
Inspiring generations

Heinemann Educational Publishers
Halley Court, Jordan Hill, Oxford OX2 8EJ
Part of Harcourt Education

Heinemann is the registered trademark of
Harcourt Education Limited

First published 2005

08 07 06 05
10 9 8 7 6 5 4 3 2 1

British Library Cataloguing in Publication Data is available from the British
Library on request.

10-digit ISBN: 0 435 513567
13-digit ISBN: 978 0 435 513566

Edited by Alex Sharpe, Standard Eight Limited
Typeset and illustrated by Tech-Set Limited, Gateshead, Tyne & Wear
Original illustrations © Harcourt Education Limited, 2005
Cover design by mccdesign ltd
Printed in the United Kingdom by Scotprint.

About this book

This book is designed to help you get your best possible grade in your Pure Core Maths 1 examination. The author is a Principal examiner, and has a good understanding of AQA's requirements.

Revise for Core 1 covers the key topics that are tested in the Pure Core Maths 1 exam paper. You can use this book to help you revise at the end of your course, or you can use it throughout your course alongside the course textbook, *Advancing Maths for AQA AS & A level Pure Core Maths 1 & 2*, which provides complete coverage of the syllabus.

Helping you prepare for your exam

To help you prepare, each topic offers you:

- **Key points to remember** – summarise the mathematical ideas you need to know and be able to use.

- **Worked examples** – help you understand and remember important methods, and show you how to set out your answers clearly.

- **Revision exercises** – help you practise using these important methods to solve problems. Exam-level questions are included so you can be sure you are reaching the right standard, and answers are given at the back of the book so you can assess your progress.

- **Test Yourself questions** – help you see where you need extra revision and practice. If you do need extra help, they show you where to look in the *Advancing Maths for AQA AS & A level Pure Core Maths 1 & 2* textbook and which example to refer to in this book.

Exam practice and advice on revising

Examination style paper – this paper at the end of the book provides a set of questions of examination standard. It gives you an opportunity to practise taking a complete exam before you meet the real thing. The answers are given at the back of the book.

How to revise – for advice on revising before the exam, read the *How to revise* section on the next page.

How to revise using this book

Making the best use of your revision time

The topics in this book have been arranged in a logical sequence so you can work your way through them from beginning to end. But **how** you work on them depends on how much time there is between now and your examination.

If you have plenty of time before the exam then you can **work through each topic in turn**, covering the key points and worked examples before doing the revision exercises and Test Yourself questions.

If you are short of time then you can **work through the Test Yourself sections** first, to help you see which topics you need to do further work on.

However much time you have to revise, make sure you break your revision into short blocks of about 40 minutes, separated by five- or ten-minute breaks. Nobody can study effectively for hours without a break.

Using the Test Yourself sections

Each Test Yourself section provides a set of key questions. Try each question:

- If you can do it and get the correct answer, then move on to the next topic. Come back to this topic later to consolidate your knowledge and understanding by working through the key points, worked examples and revision exercises.

- If you cannot do the question, or get an incorrect answer or part answer, then work through the key points, worked examples and revision exercises before trying the Test Yourself questions again. If you need more help, the cross-references beside each Test Yourself question show you where to find relevant information in the *Advancing Maths for AQA AS & A level Pure Core Maths 1 & 2* textbook and which example in *Revise for C1* to refer to.

Reviewing the key points

Most of the key points are straightforward ideas that you can learn: try to understand each one. Imagine explaining each idea to a friend in your own words, and say it out loud as you do so. This is a better way of making the ideas stick than just reading them silently from the page.

As you work through the book, remember to go back over key points from earlier topics at least once a week. This will help you to remember them in the exam.

CHAPTER 1

Advancing from GCSE maths: algebra review

Key points to remember

1 A linear equation is of the form $ax + b = 0$.

2 To solve a linear equation, collect all the terms involving x on one side of the equation.

3 If equations involve fractions try multiplying through by the lowest common multiple (LCM) of the denominators.

4 Simultaneous linear equations can be solved by elimination or by substitution.

5 Inequalities can be handled like equations, but remember to reverse the inequality sign whenever you multiply or divide by a negative number.

6 A function f has a defining rule such as $f(x) = x^2 + 3$ to enable you to find its value for different values of x.

Worked example 1

Solve the equation $4(2t + 3) - 2(t - 1) - 4 = 3t - (t + 2)$.

$$4(2t + 3) - 2(t - 1) - 4 = 3t - (t + 2)$$

$$\Rightarrow \quad 8t + 12 - 2t + 2 - 4 = 3t - t - 2$$

Multiplying out brackets.

$$\Rightarrow \quad 8t - 2t - 3t + t = -2 - 12 - 2 + 4$$

Using **2**

$$\Rightarrow \quad 4t = -12$$

Simplifying both sides.

$$\Rightarrow \quad t = -3$$

Dividing both sides by 4.

Worked example 2

Solve the equation $\dfrac{2(x + 2)}{3} - \dfrac{2x + 3}{4} = \dfrac{1}{2}$.

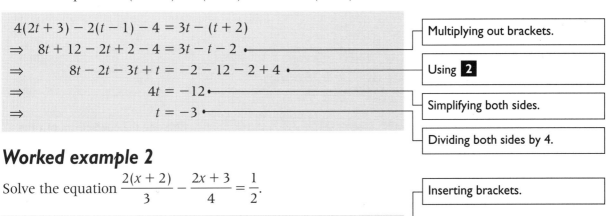

Inserting brackets.

$$\dfrac{2(x + 2)}{3} - \dfrac{(2x + 3)}{4} = \dfrac{1}{2}$$

Using **3**, multiply throughout by 12.

$$\Rightarrow \quad \dfrac{12 \times 2(x + 2)}{3} - \dfrac{12 \times (2x + 3)}{4} = \dfrac{12 \times 1}{2}$$

Cancelling

$$\Rightarrow \quad 8(x + 2) - 3(2x + 3) = 6$$

Multiplying out brackets.

$$\Rightarrow \quad 8x + 16 - 6x - 9 = 6$$

$$\Rightarrow \quad 8x - 6x = 6 - 16 + 9$$

Using **2**

$$\Rightarrow \quad 2x = -1$$

Simplifying both sides.

$$\Rightarrow \quad x = -\dfrac{1}{2}$$

Dividing both sides by 2.

Worked example 3

Solve the equation $\dfrac{80}{x+2} = \dfrac{93}{x+3}$.

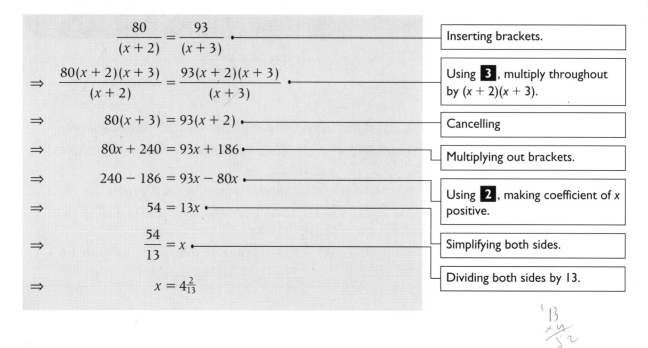

$$\frac{80}{(x+2)} = \frac{93}{(x+3)}$$ — Inserting brackets.

$$\Rightarrow \frac{80(x+2)(x+3)}{(x+2)} = \frac{93(x+2)(x+3)}{(x+3)}$$ — Using **3**, multiply throughout by $(x+2)(x+3)$.

$$\Rightarrow \quad 80(x+3) = 93(x+2)$$ — Cancelling

$$\Rightarrow \quad 80x + 240 = 93x + 186$$ — Multiplying out brackets.

$$\Rightarrow \quad 240 - 186 = 93x - 80x$$ — Using **2**, making coefficient of x positive.

$$\Rightarrow \quad 54 = 13x$$ — Simplifying both sides.

$$\Rightarrow \quad \frac{54}{13} = x$$

$$\Rightarrow \quad x = 4\tfrac{2}{13}$$ — Dividing both sides by 13.

Worked example 4

Solve the simultaneous equations: $8x - 3y = 13$
$2x - 5y = 16$

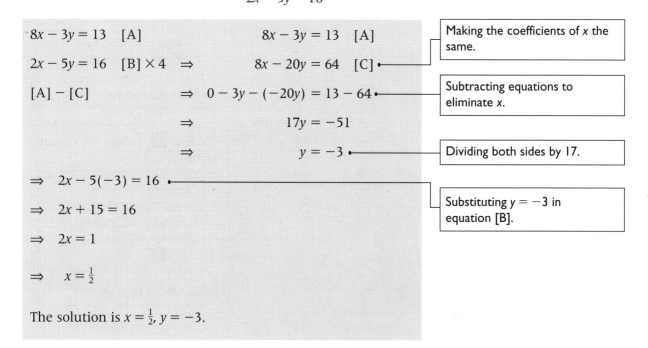

$8x - 3y = 13$ [A] $8x - 3y = 13$ [A]

$2x - 5y = 16$ [B] $\times 4 \Rightarrow$ $8x - 20y = 64$ [C] — Making the coefficients of x the same.

$[A] - [C] \qquad\qquad \Rightarrow \; 0 - 3y - (-20y) = 13 - 64$ — Subtracting equations to eliminate x.

$$\Rightarrow \qquad\qquad 17y = -51$$

$$\Rightarrow \qquad\qquad\quad y = -3$$ — Dividing both sides by 17.

$\Rightarrow \; 2x - 5(-3) = 16$ — Substituting $y = -3$ in equation [B].

$\Rightarrow \; 2x + 15 = 16$

$\Rightarrow \; 2x = 1$

$\Rightarrow \quad x = \tfrac{1}{2}$

The solution is $x = \tfrac{1}{2}$, $y = -3$.

Worked example 5

Solve the inequality $x - 3(x + 1) \geqslant 7$.

$\Rightarrow \quad x - 3x - 3 \geqslant 7$ — Multiplying out brackets.

$\Rightarrow \quad x - 3x \geqslant 7 + 3$ — Using **2**

$\Rightarrow \quad -2x \geqslant 10$ — Simplifying both sides.

$\Rightarrow \quad x \leqslant \dfrac{10}{-2}$ — Using **5**

$\Rightarrow \quad x \leqslant -5$

> You may prefer not to have to apply Key Point **5**.
> You can do this by making the coefficient of x positive,
> $-3 - 7 \geqslant 3x - x \Rightarrow \dfrac{-10}{2} \geqslant x$, which can be restated as
> $x \leqslant -5$.

Check: e.g. $x = -6$: $\quad -6 - 3(-6 + 1) = -6 - 3(-5)$
$$= -6 + 15 = 9 \geqslant 7 \quad ✓$$

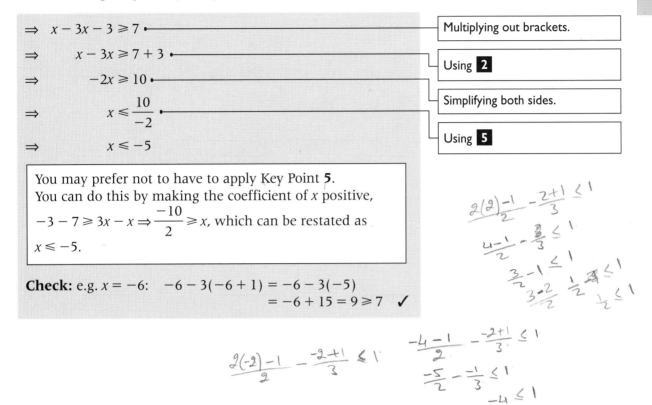

[handwritten:]
$\dfrac{2(2)-1}{2} - \dfrac{-2+1}{3} \leqslant 1$

$\dfrac{4-1}{2} - \dfrac{-3}{3} \leqslant 1$

$\dfrac{3-1}{2} \leqslant 1$

$\dfrac{3-2}{2} \quad \dfrac{1}{2} \leqslant 1 \quad \dfrac{1}{2} \leqslant 1$

$\dfrac{2(-2)-1}{2} - \dfrac{-2+1}{3} \leqslant 1$

$\dfrac{-4-1}{2} - \dfrac{-2+1}{3} \leqslant 1$

$\dfrac{-5}{2} - \dfrac{-1}{3} \leqslant 1$

$-4 \leqslant 1$

Worked example 6

Solve the inequality $\dfrac{2x - 1}{2} - \dfrac{x + 1}{3} \leqslant 1$.

[handwritten:] $\dfrac{3 \times (2x-1) - 2(x+1)}{6} \leqslant \dfrac{6 \times 1}{6}$

$\dfrac{(2x - 1)}{2} - \dfrac{(x + 1)}{3} \leqslant 1$ — Inserting brackets.

$\Rightarrow \quad \dfrac{6 \times (2x - 1)}{2} - \dfrac{6 \times (x + 1)}{3} \leqslant 6 \times 1$ — Using **3**, multiply throughout by 6.

$\Rightarrow \quad 3(2x - 1) - 2(x + 1) \leqslant 6$ — Cancelling

$\Rightarrow \quad 6x - 3 - 2x - 2 \leqslant 6$ — Multiplying out brackets.

$\Rightarrow \quad 6x - 2x \leqslant 6 + 3 + 2$

$\Rightarrow \quad 4x \leqslant 11$ — Simplifying both sides.

$\Rightarrow \quad x \leqslant \dfrac{11}{4}$ — Dividing both sides by 4.

$\Rightarrow \quad x \leqslant 2\tfrac{3}{4}$ *[handwritten: an error of publishing. -2 -]*

Check: e.g. $x = -2$: $\quad \dfrac{2(2) - 1}{2} - \dfrac{-2 + 1}{3} = \dfrac{3}{2} - 1 = \dfrac{1}{2} \leqslant 1 \quad ✓$

[handwritten: but even with -2 is good]

Worked example 7

The functions f and g are defined by $f(x) = 3(x + 2)^2$ and $g(x) = \dfrac{1}{x + 2}$, respectively. Find the values of:

(a) $f(1) - g(0)$ **(b)** $f(-1) + g(-1)$ **(c)** $f\left(-\dfrac{5}{2}\right) + g\left(-\dfrac{3}{2}\right)$

(a) $$f(1) = 3 \times (1 + 2)^2 = 3 \times (3)^2 = 3 \times 9 = 27$$

$$g(0) = \frac{1}{0 + 2} = \frac{1}{2}$$

$$\Rightarrow \qquad f(1) - g(0) = 27 - \frac{1}{2} = 26\frac{1}{2}$$

(b) $$f(-1) = 3 \times (-1 + 2)^2 = 3 \times 1 = 3$$

$$g(-1) = \frac{1}{-1 + 2} = 1$$

$$\Rightarrow \qquad f(-1) + g(-1) = 3 + 1 = 4$$

(c) $$f\left(-\frac{5}{2}\right) = 3 \times \left(-\frac{5}{2} + 2\right)^2 = 3 \times \left(-\frac{1}{2}\right)^2 = 3 \times \frac{1}{4} = \frac{3}{4}$$

$$g\left(-\frac{3}{2}\right) = \frac{1}{-\dfrac{3}{2} + 2} = \frac{1}{\dfrac{1}{2}} = 2$$

$$\Rightarrow \qquad f\left(-\frac{5}{2}\right) + g\left(-\frac{3}{2}\right) = \frac{3}{4} + 2 = 2\frac{3}{4}$$

$-\frac{5}{2} + \frac{2}{1} = \frac{-5+4}{2} = \left(\frac{-1}{2}\right)^2 = \frac{1}{4}$

$3 \times \frac{1}{4} = \frac{3}{4}$

$-\frac{3}{2} + \frac{2}{1} \quad \frac{-3+4}{2} = \frac{1}{2} \qquad \frac{1}{\frac{1}{2}}$

$\frac{1}{1} \times \frac{2}{1} = 2$

REVISION EXERCISE 1

Solve each of the equations in questions 1–12.

1 $2x - 3 = 3(1 - x)$

2 $4t - 3 - t + 7 = 3(1 - t)$

3 $4 - 3(2 - y) + 4y = 2(3 + y) - 1$

4 $2(3x - 2) - (4 - x) = 7 - 3(2 - x)$

5 $\dfrac{t}{2} + 1 = \dfrac{2t}{3} + 2$ **6** $\dfrac{1}{2}(4y - 3) = \dfrac{y}{4} + 2$

7 $\dfrac{2x - 3}{4} = \dfrac{3 - 2x}{6}$ **8** $\dfrac{4x + 5}{3} = \dfrac{x - 1}{2} + 1$

9 $\dfrac{4 - 2x}{3} + \dfrac{x + 1}{2} + \dfrac{1}{6} = 0$ **10** $\dfrac{2}{x} = \dfrac{3}{x + 1}$

11 $\dfrac{3}{x - 3} = \dfrac{2}{2 - x}$ **12** $\dfrac{1}{2x + 7} = \dfrac{4}{7(x + 2)}$

Solve the simultaneous equations in questions 13–20.

13 $x + y = 12$
      $x - y = 8$

14 $x + 2y = 9$
      $x + y = -1$

15 $2x + y = 23$
      $3x - 2y = 3$

16 $2x + 7y = 20$
      $6x + 5y = 12$

17 $y = 2x + 3$
      $y = 4 - 3x$

18 $y = 4 + x$
      $3x - 4y = 17$

19 $2a + 5b = -5$
      $4a + 3b = -17$

20 $2a + 3b = 19$
      $7a - 5b = -11$

Solve the inequalities in questions 21–27.

21 $2 + x > 4 + 3x$ **22** $2(x + 3) < 3(x - 1)$

23 $2(3x - 4) \leqslant 1 - 2x$ **24** $\dfrac{x}{2} - \dfrac{2x}{3} + 1 < 0$

25 $\dfrac{4x + 3}{5} \geqslant \dfrac{2(1 + 3x)}{7}$ **26** $\dfrac{3x - 1}{5} - \dfrac{x + 1}{3} \leqslant 1$

27 $\dfrac{5y + 2}{6} - \dfrac{4y + 3}{5} - 1 < 0$

28 The function f is defined by $f(x) = x^2 - 2x$.
Find the values of:

   (a) $f(1)$ **(b)** $f(0)$ **(c)** $f(-2)$ **(d)** $f(\tfrac{1}{2})$

29 The function g is defined by $g(x) = 2x^2 - \dfrac{3}{x}$.
Find the values of:

   (a) $g(1)$ **(b)** $g(-2)$ **(c)** $g(\tfrac{1}{2})$ **(d)** $g(-\tfrac{1}{2})$

30 The functions f and g are defined by $f(x) = (x + 2)^2$ and
$g(x) = 3x - \dfrac{1}{x^2}$ respectively. Find the values of:

   (a) $f(1) + g(1)$ **(b)** $f(-1) + g(\tfrac{1}{3})$

   (c) $g(\tfrac{1}{2}) - f(-3)$ **(d)** $g(-1) + f(-\tfrac{3}{2})$

Test yourself	What to review
	If your answer is incorrect:
1 Solve the equation $3 - 4(y + 2) = y + 5$.	See p 1 Examples 1 & 2 or review Advancing Maths for AQA C1C2 pp 2–3
2 Solve the simultaneous equations: $y = 3x + 7$ $\qquad y = 2x - 4$	See p 2 Example 4 or review Advancing Maths for AQA C1C2 p 6
3 Solve the inequality $\dfrac{x + 1}{4} - \dfrac{1 + 2x}{3} \geq 1$.	See p 3 Examples 5 & 6 or review Advancing Maths for AQA C1C2 pp 7–8
4 The function f is defined by $f(x) = 2x^2 - \dfrac{1}{x}$. Find the value of: **(a)** $f(1)$ **(b)** $f(-1)$ **(c)** $f\left(\dfrac{1}{2}\right)$ **(d)** $f\left(-\dfrac{1}{2}\right)$	See p 4 Example 7 or review Advancing Maths for AQA C1C2 p 9

Test yourself ANSWERS

4 (a) 1 **(b)** 3 **(c)** $-1\frac{1}{2}$ **(d)** $2\frac{1}{2}$

3 $x \geq -2\frac{3}{5}$

2 $x = -11,\ y = -26$

1 $y = -2$

CHAPTER 2

Surds

Key points to remember

1 A rational number is one which can be written in the form $\dfrac{a}{b}$, where a and b are integers.

2 A real number that is not rational is called irrational. Its decimal representation neither terminates nor has a recurring pattern of digits.

3 A surd is an irrational number containing a root sign.

4 $\sqrt{ab} = \sqrt{a} \times \sqrt{b}$

5 $\sqrt{\dfrac{a}{b}} = \dfrac{\sqrt{a}}{\sqrt{b}}$

6 'Like' surds can be collected. 'Unlike' surds cannot be collected.

7 To rationalise the denominator of the form $\sqrt{a}$, multiply top and bottom by $\sqrt{a}$.

8 To rationalise the denominator of the form $a + \sqrt{b}$, multiply top and bottom by $a - \sqrt{b}$.

9 To rationalise the denominator of the form $a - \sqrt{b}$, multiply top and bottom by $a + \sqrt{b}$.

Worked example 1

(a) Express each of the following in the form $n\sqrt{3}$, where n is an integer:

    (i) $\sqrt{12}$ **(ii)** $\dfrac{12}{\sqrt{3}}$

(b) Hence write $\dfrac{12}{\sqrt{3}} - \sqrt{12}$ in the form $k\sqrt{3}$, where k is an integer.

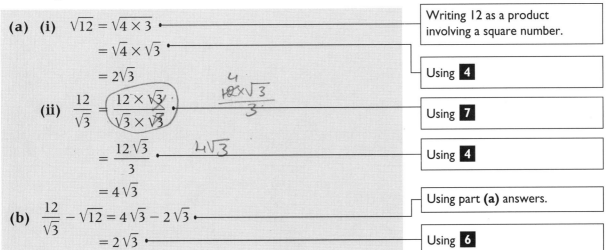

(a) (i) $\sqrt{12} = \sqrt{4 \times 3}$ — Writing 12 as a product involving a square number.

$= \sqrt{4} \times \sqrt{3}$ — Using **4**

$= 2\sqrt{3}$

(ii) $\dfrac{12}{\sqrt{3}} = \dfrac{12 \times \sqrt{3}}{\sqrt{3} \times \sqrt{3}}$ — Using **7**

$= \dfrac{12\sqrt{3}}{3}$ — Using **4**

$= 4\sqrt{3}$

(b) $\dfrac{12}{\sqrt{3}} - \sqrt{12} = 4\sqrt{3} - 2\sqrt{3}$ — Using part **(a)** answers.

$= 2\sqrt{3}$ — Using **6**

Worked example 2

(a) Simplify each of the following: **(i)** $\dfrac{\sqrt{18}}{\sqrt{2}}$ **(ii)** $\dfrac{\sqrt{245}}{\sqrt{5}}$

(b) Hence find the value of $\dfrac{\sqrt{18}}{\sqrt{2}} \times \dfrac{\sqrt{245}}{\sqrt{5}}$.

(a) **(i)** $\dfrac{\sqrt{18}}{\sqrt{2}} = \sqrt{\dfrac{18}{2}}$ — Using **5**

$= \sqrt{9}$

$= 3$

(ii) $\dfrac{\sqrt{245}}{\sqrt{5}} = \sqrt{\dfrac{245}{5}}$ — Using **5**

$= \sqrt{49}$

$= 7$

(b) $\dfrac{\sqrt{18}}{\sqrt{2}} \times \dfrac{\sqrt{245}}{\sqrt{5}} = 3 \times 7$ — Using part **(a)** answers.

$= 21$

Worked example 3

Simplify $\sqrt{20} + \sqrt{75} - \sqrt{27}$.

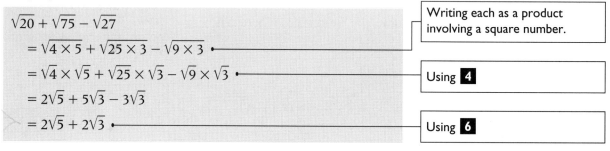

$\sqrt{20} + \sqrt{75} - \sqrt{27}$

$= \sqrt{4 \times 5} + \sqrt{25 \times 3} - \sqrt{9 \times 3}$ — Writing each as a product involving a square number.

$= \sqrt{4} \times \sqrt{5} + \sqrt{25} \times \sqrt{3} - \sqrt{9} \times \sqrt{3}$ — Using **4**

$= 2\sqrt{5} + 5\sqrt{3} - 3\sqrt{3}$

$= 2\sqrt{5} + 2\sqrt{3}$ — Using **6**

Worked example 4

Rationalise the denominator of the expression $\dfrac{2\sqrt{7}+3}{\sqrt{7}+2}$.

$$\frac{2\sqrt{7}+3}{\sqrt{7}+2} = \frac{(2\sqrt{7}+3)}{(\sqrt{7}+2)} \times \frac{(\sqrt{7}-2)}{(\sqrt{7}-2)}$$

Using **8**

$$= \frac{2\sqrt{49}-4\sqrt{7}+3\sqrt{7}-6}{\sqrt{49}-4}$$

Using **4**

$$= \frac{14-4\sqrt{7}+3\sqrt{7}-6}{7-4}$$

$$= \frac{8-\sqrt{7}}{3}$$

Using **6**

2

Worked example 5

Solve the inequality $\sqrt{2}(2+x) < 3x - 1$.

$$\sqrt{2}(2+x) < 3x - 1$$

$$\Rightarrow \quad 2\sqrt{2}+\sqrt{2}x < 3x - 1$$

Multiplying out brackets.

$$\Rightarrow \quad 2\sqrt{2}+1 < 3x - \sqrt{2}x$$

Rearranging

$$\Rightarrow \quad 2\sqrt{2}+1 < (3-\sqrt{2})x$$

Factorising

$$\Rightarrow \quad \frac{2\sqrt{2}+1}{3-\sqrt{2}} < x$$

$3 - \sqrt{2} > 0$ so equality sign does not alter.

$$x > \frac{2\sqrt{2}+1}{3-\sqrt{2}}$$

Restating

$$x > \frac{2\sqrt{2}+1}{3-\sqrt{2}} \times \frac{(3+\sqrt{2})}{(3+\sqrt{2})}$$

Using **9**

$$x > \frac{6\sqrt{2}+2\sqrt{4}+3+\sqrt{2}}{9-\sqrt{4}}$$

Using **4**

$$x > \frac{7+7\sqrt{2}}{7}$$

Using **6**

$$x > \frac{7(1+\sqrt{2})}{7}$$

Factorising

$$x > 1 + \sqrt{2}$$

Cancelling

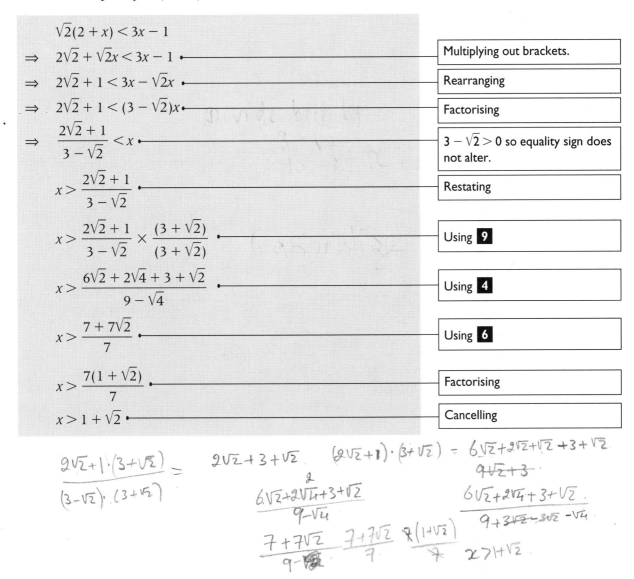

REVISION EXERCISE 2

1 Write down:

    (a) two rational numbers which lie between 2 and 3,

    (b) two irrational numbers which lie between 2 and 3.

2 Simplify the following as far as possible:

    (a) $\sqrt{28}$ **(b)** $\sqrt{63}$ **(c)** $\sqrt{32}$ **(d)** $\sqrt{150}$

3 Expand and simplify:

    (a) $\sqrt{2}(\sqrt{8} - 2\sqrt{3})$ **(b)** $\sqrt{3}(\sqrt{6} - \sqrt{27})$

    (c) $(\sqrt{5} - 1)^2$ **(d)** $(\sqrt{3} + \sqrt{6})^2$

4 Rationalise the denominator and simplify:

    (a) $\dfrac{10}{\sqrt{5}}$ **(b)** $\dfrac{2}{\sqrt{3} + 1}$ **(c)** $\dfrac{1}{\sqrt{5} - 2}$ **(d)** $\dfrac{\sqrt{2}}{\sqrt{8} - \sqrt{6}}$

5 (a) Express $(\sqrt{7} - 1)^2$ in the form $p + q\sqrt{7}$, where p and q are integers.

    (b) Hence write $\dfrac{2}{(\sqrt{7} - 1)^2}$ in the form $a + b\sqrt{7}$ and state the values of the rational numbers a and b.

6 Express each of the following in the form $p + q\sqrt{3}$, where p and q are integers.

    (a) $(7 + 3\sqrt{3})(2 - \sqrt{3})$ **(b)** $\dfrac{39}{4 - \sqrt{3}}$

7 Solve the inequality $3 < 1 - \sqrt{2}x$.

8 Rationalise the denominator and simplify $\dfrac{1 - \sqrt{2}}{2\sqrt{2} - 3}$.

9 (a) Simplify $\dfrac{(1 + \sqrt{3})(2\sqrt{3} - 5) + (1 - \sqrt{3})^2}{20}$.

    (b) Hence write $\dfrac{20}{(1 + \sqrt{3})(2\sqrt{3} - 5) + (1 - \sqrt{3})^2}$ in the form $k(1 + \sqrt{3})$ where k is an integer.

10 Simplify $\dfrac{\sqrt{6} \times \sqrt{12}}{\sqrt{2} + \sqrt{8}}$.

11 Write $\dfrac{3 - 2\sqrt{2}}{(1 + \sqrt{2})^2}$ in the form $a + b\sqrt{2}$, where a and b are integers.

12 Solve the inequality $3\sqrt{2} - 4x < -5 + 2\sqrt{2}x$.

13 (a) Simplify $(\sqrt{18} + 4)(\sqrt{18} - 4)$.

   (b) Express $\dfrac{\sqrt{18} - 4}{\sqrt{18} + 4}$ in the form $a + b\sqrt{2}$.

14 Simplify $\dfrac{39}{\sqrt{52}} - \sqrt{13}$.

15 Find the value of $\dfrac{\sqrt{300} - \sqrt{75}}{\sqrt{12} + \sqrt{3}}$.

16 (a) Express $(2\sqrt{7} - 1)(\sqrt{7} + 5)$ in the form $p(1 + \sqrt{7})$, where p is an integer.

   (b) Solve the inequality $\sqrt{7} - x < 2\sqrt{7}x - 5$.

Test yourself	**What to review**
	If your answer is incorrect:
1 State whether each of the following is rational or irrational. If rational, write it in a form not involving square roots.	See p 8 Example 2 or review Advancing Maths for AQA C1C2 p 14

(a) $\dfrac{8}{\sqrt{2}}$

(b) $\dfrac{\sqrt{8}}{\sqrt{2}}$ $\dfrac{\sqrt{4 \times 2}}{\sqrt{2}} = \dfrac{\sqrt{4}\sqrt{2}}{\sqrt{2}} = 2$

(c) $\dfrac{\sqrt{10} + \sqrt{6}}{\sqrt{2}}$

(d) $\dfrac{\sqrt{2}}{\sqrt{8} + \sqrt{18}}$ $\dfrac{\sqrt{2}}{\sqrt{4}\sqrt{2} + \sqrt{9}\sqrt{2}} = \dfrac{\sqrt{2}}{2\sqrt{2} + 3\sqrt{2}} = \dfrac{\sqrt{2}}{5\sqrt{2}} = \dfrac{1}{5}$

---

2 Simplify each of these surd expressions:	See p 8 Example 2 or review Advancing Maths for AQA C1C2 pp 15–16

(a) $\sqrt{72}$ (b) $\sqrt{125}$ (c) $\sqrt{2}\,\sqrt{3}\,\sqrt{48}$

---

3 Simplify $\sqrt{\dfrac{36}{2}} + \sqrt{50} - \sqrt{54}$.	See p 8 Example 3 or review Advancing Maths for AQA C1C2 p 18

4	Review Advancing Maths for AQA C1C2 pp 17–19

The lengths of the sides of a right-angled triangle are $6\sqrt{2}$ cm, $2\sqrt{3}$ cm and x cm. Given that $x < 6\sqrt{2}$, find the area of the triangle in the form $k\sqrt{5}$ cm².

$c^2 = a^2 + b^2$

$(6\sqrt{2})^2 = (2\sqrt{3})^2 + b^2$

$b^2 = 36 + 4$

$b^2 = 40 \qquad b = 6\sqrt{2}$

$b = \sqrt{40} \qquad \qquad 15\sqrt{5}$

5 Rationalise the denominator of each of the following. Simplify where possible.

(a) $\dfrac{14}{\sqrt{7}}$ (b) $\dfrac{2}{3-\sqrt{5}}$ (c) $\dfrac{4+\sqrt{2}}{2-\sqrt{2}}$

See p 9 Example 4 or review Advancing Maths for AQA C1C2 pp 21–22

6 Solve the inequality $2\sqrt{3}-2x<\sqrt{3}x-1$.

See p 9 Example 5 or review Advancing Maths for AQA C1C2 p 23

Test yourself ANSWERS

1 (a) irrational (b) rational = 2 (c) irrational (d) rational $=\dfrac{1}{5}$

2 (a) $6\sqrt{2}$ (b) $5\sqrt{5}$ (c) $12\sqrt{2}$

3 $8\sqrt{2}-3\sqrt{6}$

4 $6\sqrt{5}\ \text{cm}^2$

5 (a) $2\sqrt{7}$ (b) $\dfrac{1}{2}(3+\sqrt{5})$ (c) $5+3\sqrt{2}$

6 $x>3\sqrt{3}-4$

CHAPTER 3

Coordinate geometry of straight lines

3

Key points to remember

1 The distance between the points (x_1, y_1) and (x_2, y_2) is $\sqrt{(x_2 - x_1)^2 + (y_2 - y_1)^2}$.

2 The coordinates of the mid-point of the line segment joining (x_1, y_1) and (x_2, y_2) are $\left(\dfrac{x_1 + x_2}{2}, \dfrac{y_1 + y_2}{2}\right)$.

3 The gradient of a line joining the two points $A(x_1, y_1)$ and $B(x_2, y_2) = \dfrac{y_2 - y_1}{x_2 - x_1}$.

4 Lines with gradients m_1 and m_2:
 – are parallel if $m_1 = m_2$,
 – are perpendicular if $m_1 \times m_2 = -1$.

5 $y = mx + c$ is the equation of a straight line with gradient m and y-intercept c.

6 $ax + by + c = 0$ is the general equation of a line. It has gradient $-\dfrac{a}{b}$ and y-intercept $-\dfrac{c}{b}$.

7 The equation of the straight line which passes through the point (x_1, y_1) and has gradient m is $y - y_1 = m(x - x_1)$.

8 The equation of the straight line which passes through the points (x_1, y_1) and (x_2, y_2) is $\dfrac{y - y_1}{y_2 - y_1} = \dfrac{x - x_1}{x_2 - x_1}$.

9 A point lies on a line if the coordinates of the point satisfy the equation of the line.

10 Given accurately drawn graphs of the two intersecting straight lines with equations $ax + by + c = 0$ and $Ax + By + C = 0$, the coordinates of the point of intersection can be read off. These coordinates give the solution of the simultaneous equations $ax + by + c = 0$ and $Ax + By + C = 0$.

11 To find the coordinates of the point of intersection of the two lines with equations $ax + by + c = 0$ and $Ax + By + C = 0$, you solve the two equations simultaneously.

Worked example 1

The coordinates of the points A, B and C are $(1, 2)$, $(8, 3)$ and $(10, 1)$ respectively.

(a) Find the distance AB giving your answer in the form $k\sqrt{2}$, where k is an integer.

(b) The mid-point of BC is M. Find the coordinates of M.

(c) Find the area of triangle ABM.

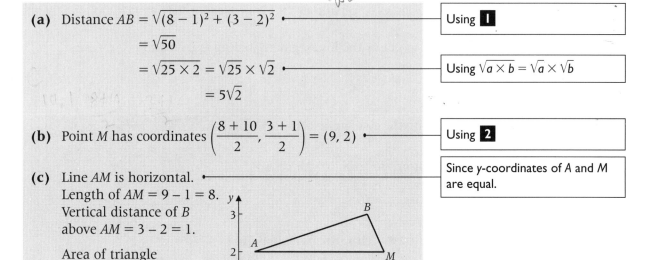

(a) Distance $AB = \sqrt{(8-1)^2 + (3-2)^2}$ • ——————— Using **1**

$\qquad = \sqrt{50}$

$\qquad = \sqrt{25 \times 2} = \sqrt{25} \times \sqrt{2}$ • ——————— Using $\sqrt{a \times b} = \sqrt{a} \times \sqrt{b}$

$\qquad = 5\sqrt{2}$

(b) Point M has coordinates $\left(\dfrac{8+10}{2}, \dfrac{3+1}{2}\right) = (9, 2)$ • ——————— Using **2**

(c) Line AM is horizontal. • ——————— Since y-coordinates of A and M are equal.

Length of $AM = 9 - 1 = 8$.
Vertical distance of B above $AM = 3 - 2 = 1$.

Area of triangle
$ABM = \dfrac{1}{2} \times 8 \times 1 = 4$.

Worked example 2

(a) Find the gradient of the line joining the points $A(2, 5)$ and $B(1, 2)$.

(b) Find the equation of the line which is parallel to AB and cuts the y-axis at the point $C(0, 7)$.

(c) Show that the line with equation $3y + x = 21$ passes through C and is perpendicular to AB.

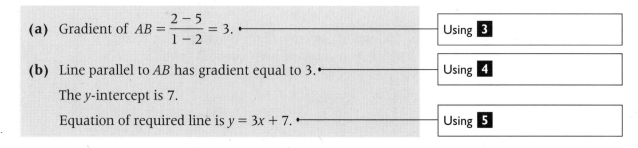

(a) Gradient of $AB = \dfrac{2-5}{1-2} = 3$. • ——————— Using **3**

(b) Line parallel to AB has gradient equal to 3. • ——————— Using **4**

The y-intercept is 7.

Equation of required line is $y = 3x + 7$. • ——————— Using **5**

(c) Since $3 \times 7 + 0 = 21$, the point $C(0, 7)$ lies on the line $3y + x = 21$.

| Using **9** |

$$3y + x = 21 \quad \Rightarrow \quad 3y = -x + 21 \quad \Rightarrow \quad y = -\frac{1}{3}x + 7$$

$$\Rightarrow \quad \text{gradient} = -\frac{1}{3}$$

| Using **5** |

Since $-\frac{1}{3} \times 3 = -1$, the lines $3y + x = 21$ and AB are perpendicular.

| Using **4** |

3

So the line $3y + x = 21$ passes through C and is perpendicular to AB.

Worked example 3

(a) Find an equation of the straight line L, which passes through the point $A(1, -3)$ and is perpendicular to the line with equation $3x + 2y = 0$.

| 'Find an equation…' means any correct form of the equation of the line will be acceptable. |

(b) The points C and D have coordinates $(8, -10)$ and $(-4, -4)$ respectively. Find an equation of the straight line CD.

(c) The line CD intersects the straight line L at the point R. Find the coordinates of R.

(a) The gradient of $3x + 2y = 0$ is $-\frac{3}{2}$.

| Using **6** |

$$-\frac{3}{2} \times m_2 = -1 \quad \Rightarrow \quad \text{gradient of line } L \text{ is } \frac{2}{3}$$

| Using **4** |

Equation of line L is $y - (-3) = \frac{2}{3}(x - 1)$

| Using **7** |

$$\Rightarrow \quad y + 3 = \frac{2}{3}(x - 1)$$

or $\quad 3y = 2x - 11 \quad$ **[I]**

(b) Equation of line CD is $\dfrac{y - (-10)}{-4 - (-10)} = \dfrac{x - 8}{-4 - 8}$

| Using **8** |

$$\Rightarrow \quad \frac{y + 10}{6} = \frac{x - 8}{-12}$$

or $\quad 2y + x + 12 = 0 \quad$ **[II]**

(c) $2y + x + 12 = 0 \quad \Rightarrow \quad x = -2y - 12$

Substitute in [I] $\quad \Rightarrow \quad 3y = 2(-2y - 12) - 11$

| Using **11** |

$$\Rightarrow \quad 7y = -35 \quad \Rightarrow \quad y = -5$$

When $y = -5, x = -2(-5) - 12 = -2$.

R has coordinates $(-2, -5)$.

Worked example 4

The points A, B, C and D have coordinates $(0, 3)$, $(0, -2)$, $(6, 4)$ and $(3, 6)$ respectively.

(a) Find, in the form $y = mx + c$, the equations of the straight lines AD and BC.

(b) Explain why $ABCD$ is a trapezium.

(c) The perpendicular from A to BC intersects BC at the point R.
  (i) Find the coordinates of R.
  (ii) Find the length of AR giving your answer in the form $k\sqrt{2}$, where k is a rational number.

(d) Calculate the area of $ABCD$.

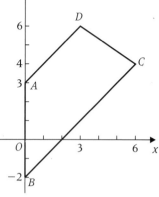

(a) Equation of AD: $\dfrac{y-3}{6-3} = \dfrac{x-0}{3-0}$ or $y = x + 3$

> Using **8**

   Equation of BC: $\dfrac{y-(-2)}{4-(-2)} = \dfrac{x-0}{6-0}$ or $y = x - 2$

> Using **8**

(b) $m_{AD} = 1$ and $m_{BC} = 1$

> Using **5**

$\Rightarrow$ AD is parallel to BC

AB is vertical, DC is not vertical so AB and DC are not parallel.

> Using **4**

$ABCD$ is therefore a trapezium with parallel sides AD and BC.

(c) **(i)** $m_{BC} \times m_{AR} = -1$ $\Rightarrow$ $m_{AR} = -1$

> Using **4**

Equation of AR is $y = -x + 3$.

R is the point of intersection of $y = -x + 3$ and $y = x - 2$.

> Using **5**

$\Rightarrow$ $x - 2 = -x + 3$

> Using **11**

$\Rightarrow$ $2x = 5$ $\Rightarrow$ $x = \dfrac{5}{2}$.

When $x = \dfrac{5}{2}$, $y = \dfrac{5}{2} - 2 = \dfrac{1}{2}$.

$\Rightarrow$ R is $\left(\dfrac{5}{2}, \dfrac{1}{2}\right)$

(ii) $AR = \sqrt{\left(\dfrac{5}{2} - 0\right)^2 + \left(\dfrac{1}{2} - 3\right)^2}$

> Using **1**

$AR = \sqrt{\dfrac{25}{4} + \dfrac{25}{4}} = \sqrt{\dfrac{25}{4} \times 2} = \sqrt{\dfrac{25}{4}} \times \sqrt{2}$

> Using $\sqrt{a \times b} = \sqrt{a} \times \sqrt{b}$.

$AR = \dfrac{5}{2}\sqrt{2}$

> Using **1**

(d) $AD = \sqrt{3^2 + 3^2} = \sqrt{9 \times 2} = 3\sqrt{2}$

$BC = \sqrt{6^2 + 6^2} = \sqrt{36 \times 2} = 6\sqrt{2}$

> Using **1**

Area of trapezium $ABCD = \dfrac{1}{2}(3\sqrt{2} + 6\sqrt{2}) \times \dfrac{5}{2}\sqrt{2}$

> Area of trapezium $= \dfrac{1}{2}(a + b)h$.

$= \dfrac{90}{4} = 22\dfrac{1}{2}$

REVISION EXERCISE 3

1 Find the distance between the points $(-1, -4)$ and $(5, 4)$.

2 (a) Find the coordinates of the mid-point M of the line segment PQ, where P is the point $(3, 1)$ and Q is the point $(7, -3)$.

(b) Find the distance OM, where O is the origin.

3 (a) Find the gradient of the line joining the points $A(-2, 4)$ and $B(3, -6)$.

(b) Find the gradient of the line joining the points $P(-1, -1)$ and $Q(3, 1)$.

(c) What can you deduce about the lines AB and PQ?

4 (a) Write down an equation of the straight line, L, which has gradient 4 and y-intercept 5.

(b) Find an equation of the straight line which passes through the point $(-2, 0)$ and is parallel to the line L.

5 (a) Find an equation of the line passing through the points $(-4, 2)$ and $(1, -8)$.

(b) This line intersects the line $y = x$ at the point R. Find the coordinates of R.

6 (a) Show that the point $A(-1, 3)$ lies on each of the straight lines with equations $x + y = 2$ and $4x + 3y = 5$.

(b) Find the equation of the straight line which joins the origin to the point of intersection of the lines $x + y = 2$ and $4x + 3y = 5$.

7 The point A has coordinates $(-2, 4)$ and the point B has coordinates $(2, -2)$.

(a) Find the distance AB, giving your answer in the form $k\sqrt{13}$, where k is an integer.

(b) Show that the mid-point of AB lies on the y-axis.

(c) Find the gradient of AB.

8 The line PQ, with equation $2y = 4x - 3$, is parallel to the line SR, where S has coordinates $(-2, 2)$ and R has coordinates $(4, k)$.

(a) Find the value of the constant k.

(b) Given that P has coordinates $(1, p)$, find the value of p.

(c) Show that PS is perpendicular to PQ.

3

9 The straight lines with equations $2x - 3y = 6$ and $4x + 5y = 1$ intersect at the point P.

  (a) Find the coordinates of P.

  (b) The point S has coordinates $\left(-\dfrac{1}{2}, 3\right)$. Find the equation of the straight line PS, giving your answer in the form $y = ax + b$.

  (c) Find an equation of the line PQ, given that PQ is perpendicular to PS.

10 The points A and B have coordinates $(-1, 2)$ and $(3, 10)$, respectively. The mid-point of the line segment AB is M.

  (a) Find the coordinates of M.

  (b) Find the distance AM, giving your answer in the form $k\sqrt{5}$.

11 Show that the straight line through the points $(-5, 1)$ and $(1, -7)$ is perpendicular to the line with equation $6x - 8y + 1 = 0$.

12 Find, in the form $y = ax + b$, the equation of the straight line through the point $(0, 4)$ parallel to the line with equation $5x - 2y = 3$.

13 Find an equation of the perpendicular bisector of the line segment PQ, where P is the point $(2, -1)$ and Q is the point $(-4, 7)$.

14 Show that $A(-6, -7)$, $B(2, 5)$, and $C(4, 8)$ are collinear points.

15 Find the coordinates of the points of intersection of the straight lines with equations $y = x + 3$ and $3y = 2x - 4$.

16 Show that the point of intersection of the straight lines with equations $2x - y + 4 = 0$ and $5y - 7x = 20$ is the y-intercept of the straight line with equation $2y - 3x = 8$.

17 The perpendicular bisector of the line joining the points $P(-1, 3)$ and $Q(3, 2)$ meets the x-axis at the point A and the y-axis at the point B. Find the area of triangle OAB, where O is the origin.

18 The point $A(3k, k)$ lies on the line with equation $3x + y = 7$.

  (a) Find the value of k.

  (b) Find the gradient of OA, where O is the origin.

  (c) Find the length of the perpendicular from the origin to the line $3x + y = 7$.

19 (a) Find an equation of the perpendicular from the origin to the straight line with equation $3x + 4y - 1 = 0$.

  (b) Find the coordinates of the point where this perpendicular intersects the line $3x + 4y - 1 = 0$.

  (c) Hence find the perpendicular distance from the origin to the line $3x + 4y - 1 = 0$.

Test yourself	**What to review**
	If your answer is incorrect:

1 The mid-point of the line segment joining $P(-1, 4)$ and $Q(5, 0)$ is M. The point R has coordinates $(0, -1)$.
   (a) Show that the length of PQ is $2\sqrt{13}$.
   (b) Show that RM is perpendicular to PQ.
   (c) **(i)** Find the length of RM leaving your answer in surd form.
       (ii) Hence find the size of angle PRM.
	See p 14 Example 1 or review Advancing Maths for AQA C1C2 pp 30–31, 37

2 Find an equation of the straight line passing through the point $(1, 3)$ which is parallel to the line with equation $4x - 2y + 7 = 0$.	See p 15 Example 3 or review Advancing Maths for AQA C1C2 p 40

3 The points A and B have coordinates $(-2, 3)$ and $(1, -3)$, respectively.
   (a) Find the equation of the line AB, giving your answer in the form $y = mx + c$.
   (b) The line AB intersects the line with equation $3y - 4x = 7$ at the point P. Find the coordinates of P.
	See p 14 Example 2 or review Advancing Maths for AQA C1C2 p 42–43

4 **(a)** Find the gradient and the y-intercept of the straight line with equation $3x - 4y = 12$.
   (b) The point $T(2k, -3k)$ lies on the line $3x - 4y = 12$. Find the value of k.
   (c) Find the gradient of the line OT, where O is the origin.
   (d) The line $3x - 4y = 12$ intersects the x-axis at the point S. Find the area of triangle OST.
	See p 14 Examples 1 & 2 or review Advancing Maths for AQA C1C2 pp 38–39

3

Test yourself ANSWERS

4 (a) Gradient $= \frac{3}{4}$; y-intercept is -3 **(b)** $\frac{2}{3}$ **(c)** $-\frac{3}{2}$ **(d)** 4

3 (a) $y = -2x - 1$ **(b)** $(-1, 1)$

2 $2x - y + 1 = 0$

1 (c) (i) $\sqrt{13}$ **(ii)** $45°$

CHAPTER 4

Quadratic functions and their graphs

Key points to remember

1 An expression of the form $ax^2 + bx + c$ is a quadratic and the graph of $y = ax^2 + bx + c$ is called a parabola.

2 A quadratic equation that can be written in the form $(x - p)(x - q) = 0$ has roots p and q. The graph of $y = (x - p)(x - q)$ crosses the x-axis at the points $(p, 0)$ and $(q, 0)$.

3 A quadratic can be written in the form $A(x + B)^2 + C$ and this is called the completed square form.
This form enables you to find the greatest or least values of the quadratic.

4 In general, a translation of $\begin{bmatrix} a \\ 0 \end{bmatrix}$ transforms the graph of $y = \mathrm{f}(x)$ into the graph of $y = \mathrm{f}(x - a)$.

5 In general, a translation of $\begin{bmatrix} a \\ b \end{bmatrix}$ transforms the graph of $y = \mathrm{f}(x)$ into the graph of $y - b = \mathrm{f}(x - a)$.

6 The formula $x = \dfrac{-b \pm \sqrt{b^2 - 4ac}}{2a}$ can be used to find the solutions to any quadratic equation and this formula must be learned off by heart.

7 The expression $b^2 - 4ac$ is called the discriminant.
When $b^2 - 4ac$ is a perfect square, the roots of the quadratic equation are rational and the quadratic will factorise.
When $b^2 - 4ac > 0$, the quadratic equation has two distinct real roots.
When $b^2 - 4ac = 0$, the quadratic equation has one (repeated) root. This condition is sometimes called having equal roots.
When $b^2 - 4ac < 0$, the quadratic equation has no real roots.

Worked example 1

(a) Find the coordinates of the points where the parabola with equation $y = 2x^2 - 5x - 12$ intersects the coordinate axes.

(b) Find the equation of the line of symmetry of the parabola.

(a) $y = 2x^2 - 5x - 12$

$\quad x = 0 \implies y = 0 - 0 - 12 = -12$ — On y-axis, $x = 0$.

$\quad y = 0 \implies 2x^2 - 5x - 12 = 0$ — On x-axis, $y = 0$.

$\quad\quad \implies (2x + 3)(x - 4) = 0$ — Factorising

$\quad\quad \implies x = -\dfrac{3}{2}, x = 4$ — $p \times q = 0 \implies p = 0$ or $q = 0$

Parabola intersects coordinate axes at

$(0, -12), \left(-\dfrac{3}{2}, 0\right), (4, 0)$.

(b) Mid-point of line joining $\left(-\dfrac{3}{2}, 0\right)$ and $(4, 0)$

is $\left(\dfrac{-\dfrac{3}{2} + 4}{2}, \dfrac{0 + 0}{2}\right) = \left(\dfrac{5}{4}, 0\right)$. — Axis of symmetry is perpendicular bisector of line joining points where parabola intersects x-axis.

Equation of line of symmetry is $x = \dfrac{5}{4}$.

4

Worked example 2

A curve has equation $y = x^2 + 6x + 11$.

(a) Express the quadratic $x^2 + 6x + 11$ in the form $(x + p)^2 + q$, where p and q are integers.

(b) Hence, or otherwise:
   (i) show that the curve does not intersect the x-axis,
   (ii) find the coordinates of the minimum point.

(a) $\quad (x + 3)^2 = x^2 + 6x + 9$ — 3 is half of the coefficient of x in $x^2 + 6x + 11$.

$\quad x^2 + 6x + 11 = x^2 + 6x + 9 - 9 + 11$ — Completing the square.

$\quad\quad = (x + 3)^2 + 2$

(b) (i) $y = (x + 3)^2 + 2$

Intersects x-axis if $(x + 3)^2 + 2 = 0$ — On x-axis, $y = 0$.

$\implies (x + 3)^2 = -2$, which is impossible, so curve does not intersect x-axis. — $(x + 3)^2 \geqslant 0$ for real x.

(ii) $y = (x + 3)^2 + 2$

Since $(x + 3)^2 \geqslant 0$, the minimum value of y is 2 and occurs when $(x + 3)^2 = 0 \implies x = -3$.

$(-3, 2)$ is the minimum point.

Worked example 3

A parabola has equation $y = 3 + 4x - x^2$.

(a) Express $3 + 4x - x^2$ in the form $a - (x - b)^2$, where a and b are integers.

(b) Hence find the coordinates of the vertex of the parabola.

(c) **(i)** Find, in an exact form, the coordinates of the points where the parabola crosses the coordinate axes.
   (ii) Sketch the graph of the parabola.

(d) State the geometrical transformation which maps the graph of $y = -x^2$ onto the graph of $y = 3 + 4x - x^2$.

(a) $3 + 4x - x^2 \equiv a - (x - b)^2$

$\equiv a - (x^2 - 2bx + b^2)$

$3 + 4x - x^2 \equiv a - b^2 + 2bx - x^2$

$\Rightarrow \quad 4 = 2b \quad \Rightarrow \quad b = 2$ ———— Comparing coefficients of x.

$\Rightarrow \quad 3 = a - b^2$ ———— Comparing the constant terms.

$\Rightarrow \quad 3 = a - 4 \quad \Rightarrow \quad a = 7$.

$3 + 4x - x^2 \equiv 7 - (x - 2)^2$

(b) $y = 7 - (x - 2)^2$

The greatest value of y occurs when $x - 2 = 0 \Rightarrow x = 2$. ———— $(x - 2)^2 \geqslant 0$ for real x.

When $x = 2$, $y = 7 - 0 = 7$.

So the vertex of the parabola is $(2, 7)$.

(c) **(i)** When $x = 0$, $y = 3$

When $y = 0$, $7 - (x - 2)^2 = 0 \Rightarrow (x - 2)^2 = 7$

$\Rightarrow \quad x - 2 = \pm\sqrt{7}$

$\Rightarrow \quad x = 2 \pm \sqrt{7}$

Parabola crosses coordinate axes at
$(0, 3)$, $(2 - \sqrt{7}, 0)$, $(2 + \sqrt{7}, 0)$.

(ii)

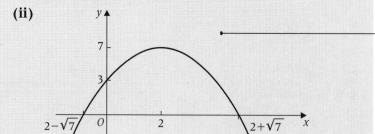

Since the coefficient of x^2 is negative, the graph is $\cap$-shaped.

(d) $y = 7 - (x - 2)^2 \quad \Rightarrow \quad y - 7 = -(x - 2)^2$

$y = -x^2$ to $y - 7 = -(x - 2)^2 \quad \Rightarrow \quad$ Translation $\begin{bmatrix} 2 \\ 7 \end{bmatrix}$. ———— Using **5**

Worked example 4

(a) By finding the discriminant of the equation
$3x^2 - 4x - 1 = 0$, state whether the equation has rational
or irrational roots.

(b) Find the roots of the equation $3x^2 - 4x = 1$, leaving your
answer in a simplified surd form.

(c) Find the possible values of k for which the equation
$3x^2 - 4x - k = 0$ has two distinct real roots.

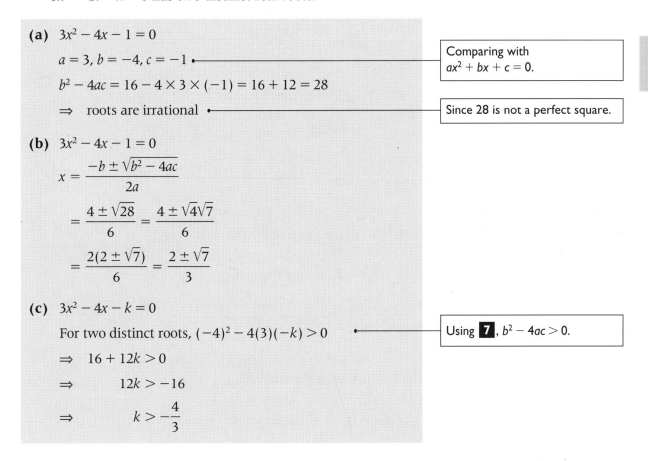

(a) $3x^2 - 4x - 1 = 0$

$a = 3, b = -4, c = -1$ — Comparing with $ax^2 + bx + c = 0$.

$b^2 - 4ac = 16 - 4 \times 3 \times (-1) = 16 + 12 = 28$

$\Rightarrow$ roots are irrational — Since 28 is not a perfect square.

(b) $3x^2 - 4x - 1 = 0$

$$x = \frac{-b \pm \sqrt{b^2 - 4ac}}{2a}$$

$$= \frac{4 \pm \sqrt{28}}{6} = \frac{4 \pm \sqrt{4}\sqrt{7}}{6}$$

$$= \frac{2(2 \pm \sqrt{7})}{6} = \frac{2 \pm \sqrt{7}}{3}$$

(c) $3x^2 - 4x - k = 0$

For two distinct roots, $(-4)^2 - 4(3)(-k) > 0$ — Using **7**, $b^2 - 4ac > 0$.

$\Rightarrow \quad 16 + 12k > 0$

$\Rightarrow \quad\quad 12k > -16$

$\Rightarrow \quad\quad\quad k > -\dfrac{4}{3}$

4

Worked example 5

The graph of $y = x^2 + 4x$ is translated by $\begin{bmatrix} 1 \\ 2 \end{bmatrix}$. Find the equation
of the new graph, giving your answer in the form
$y = ax^2 + bx + c$.

$y = x^2 + 4x$ becomes $y - 2 = (x - 1)^2 + 4(x - 1)$. — Using **5**

$\Rightarrow \quad y = x^2 - 2x + 1 + 4x - 4 + 2$

$\Rightarrow \quad y = x^2 + 2x - 1$

REVISION EXERCISE 4

1 Find the coordinates of the points where the curve with equation $y = x^2 - 6x + 5$ intersects the coordinate axes.

2 The curve $y = x^2 - 2x - 8$ cuts the x-axis at the points A and B. Calculate the distance AB.

3 (a) Find the coordinates of the points where the parabola with equation $y = 6x - 4x^2$ crosses the x-axis.

(b) Find the equation of the axis of symmetry of this parabola.

4 (a) Write $x^2 - 8x + 13$ in the form $(x - p)^2 + q$, where p and q are integers.

(b) Hence find:

(i) the vertex of the parabola $y = x^2 - 8x + 13$,

(ii) the equation of the axis of symmetry of the parabola.

5 (a) Express $3x^2 + 18x + 2$ in the form $A(x + B)^2 + C$, where A, B and C are integers.

(b) Find the least value of $3x^2 + 18x + 2$ and state the value of x at which it occurs.

6 A curve C, has equation $y = x^2 - 5x - 6$.

(a) Factorise $x^2 - 5x - 6$.

(b) Find the coordinates of the points where the curve, C, intersects the coordinate axes.

(c) The line of symmetry of the curve C has equation $x = k$. Find the value of the constant k.

7 Sketch the graph of the curve with equation $y = (x + 3)(x - 2)$ indicating the coordinates of the points where the curve crosses the coordinate axes.

8 Determine the greatest value of $f(x) = -x^2 + 4x$ and state the value of x at which it occurs.

9 (a) Express $x^2 + 4x + 5$ in the form $(x + p)^2 + q$.

(b) Hence find the coordinates of the vertex of the curve $y = x^2 + 4x + 5$.

(c) Write down the equation of the line of symmetry of the curve.

10 (a) Find the roots of $2x^2 - 4x = 1$ giving your answers in a . simplified surd form.

(b) The roots of the equation $2x^2 - 4x = k$ are real and different. Find the possible values of k.

11 Find the values of k for which the equation $kx^2 + 12x + (k - 5) = 0$ has equal roots.

12 (a) Describe the geometrical transformation that maps the graph of $y = x^2$ onto the graph of $y = (x - 2)^2$.

(b) (i) Write $x^2 - 4x + 7$ in the form $(x - a)^2 + b$.

(ii) Describe the geometrical transformation that maps the graph of $y = x^2$ onto the graph of $y = x^2 - 4x + 7$.

13 Find the equation of the graph of $y = x^2$ after it has been translated by the given vectors, giving your answer in the form $y = x^2 + px + q$:

(a) $\begin{bmatrix} -1 \\ 0 \end{bmatrix}$ **(b)** $\begin{bmatrix} 0 \\ 2 \end{bmatrix}$ **(c)** $\begin{bmatrix} 1 \\ 3 \end{bmatrix}$ **(d)** $\begin{bmatrix} -2 \\ -5 \end{bmatrix}$

14 Find the equation of the graph of $y = -x^2$ after it has been translated by the vector $\begin{bmatrix} 1 \\ 2 \end{bmatrix}$, giving your answer in the form $y = a + bx - x^2$.

15 (a) Sketch the graph of $y = (x - 2)^2$.

(b) Factorise $x^2 - 2x + 1$.

(c) Describe the geometrical transformation that maps the graph of $y = (x - 2)^2$ onto the graph of $y = x^2 - 2x + 1$.

16 The function f is defined by $f(x) = (2x + 3)(2x - 3)$.

(a) Find the coordinates of the points where the graph of $y = f(x)$ cuts the coordinate axes.

(b) Sketch the graph of $y = f(x)$.

(c) The graph of $y = f(x)$ is translated by the vector $\begin{bmatrix} 1 \\ -2 \end{bmatrix}$ to give the graph of $y = g(x)$. Find an expression for $g(x)$ in the form $ax^2 + bx + c$, where a, b and c are integers.

17 (a) Factorise $7x^2 - 12x - 64$.

(b) The equation $kx^2 - (k + 8)x + 2k + 1 = 0$ has equal roots. Find the possible values of k.

Test yourself	What to review
	If your answer is incorrect:
1 Find the coordinates of the points where the parabola $y = 4x^2 - 5x - 6$ crosses the x-axis and state the equation of the line of symmetry of the parabola.	See pp 20–21 Example 1 or review Advancing Maths for AQA C1C2 pp 52–54
2 (a) Determine the value of each of the constants p and q such that $p - (2x + q)^2 = 7 - 12x - 4x^2$. **(b)** Hence, or otherwise, determine the greatest value of $7 - 12x - 4x^2$ and state the value of x for which this greatest value occurs.	See p 21 Example 2 or review Advancing Maths for AQA C1C2 pp 55–58
3 The graph of $y = (x - 2)^2 - 4$ has been transformed from the graph of $y = x^2$. State the vector of the translation.	See p 23 Example 5 or review Advancing Maths for AQA C1C2 pp 61–62
4 Find the exact solutions of the equation $6 - 4x - x^2 = 0$, giving your answers in a simplified surd form.	See p 23 Example 4 or review Advancing Maths for AQA C1C2 p 63
5 The equation $(p - 1)x^2 - 12x + (p + 4) = 0$ has repeated roots. Find the possible values of the constant p.	See p 23 Example 4 or review Advancing Maths for AQA C1C2 pp 64–66

Test yourself ANSWERS

5 $-8, 5$

4 $-2 \mp \sqrt{10}$

3 $\begin{bmatrix} 2 \\ -4 \end{bmatrix}$

2 (a) $p = 16, q = 3$ **(b)** $16, -\dfrac{3}{2}$

1 $\left(-\dfrac{3}{4}, 0\right), (2, 0), x = \dfrac{5}{8}$

CHAPTER 5

Polynomials

Key points to remember

1 An expression of the form $ax^n + bx^{n-1} + \ldots + px^2 + qx + r$ (where $a, b, \ldots, p, q, r$ are constants) is called a polynomial in x.

2 The **degree** of a polynomial is given by the highest power of the variable.

3 Polynomials can be added or subtracted by collecting **'like'** terms.

4 Two polynomials can be multiplied by taking the second bracket and multiplying it by each term in the first bracket.

5 For the polynomial $P(x) = ax^n + bx^{n-1} + \ldots + px^2 + qx + r$, the coefficient of x^n is $a, \ldots$, the coefficient of x is q and the constant term is r.

Worked example 1

Two of the following expressions are polynomials. Multiply the two polynomials and state the degree of the polynomial in your answer:

$$\frac{1}{2}x^3 - 2x + 7 \qquad 2 + 2x^2 + 4x \qquad \frac{x^3 + x^2 - 3x}{x^4 - 7}$$

$\dfrac{x^3 + x^2 - 3x}{x^4 - 7}$ is not a polynomial. — Using **1**

$\left(\dfrac{1}{2}x^3 - 2x + 7\right)(2x^2 + 4x + 2)$ — Writing the polynomials in descending powers of x.

$\quad = \dfrac{1}{2}x^3(2x^2 + 4x + 2) - 2x(2x^2 + 4x + 2) + 7(2x^2 + 4x + 2)$ — Using **4**

$\quad = x^5 + 2x^4 + x^3 - 4x^3 - 8x^2 - 4x + 14x^2 + 28x + 14$

$\quad = x^5 + 2x^4 - 3x^3 + 6x^2 + 24x + 14$ — Using **3**

The degree of this polynomial is 5. — Using **2**

Worked example 2

The cubic polynomial $p(t)$ is given by $(t - 1)(t + 1)(t - 2)$.

(a) Show that $p(t)$ can be written in the form $t^3 + at^2 + bt + c$, where a, b and c are constants whose values are to be found.

(b) Find the possible values of k for which $p(k) = 0$.

(a) $(t-1)(t+1) = t^2 + t - t - 1 = t^2 - 1$ — Multiplying first two pairs of brackets.

$\Rightarrow (t-1)(t+1)(t-2) = (t^2 - 1)(t-2)$

$= t^2(t-2) - 1(t-2)$ — Using **4**

$= t^3 - 2t^2 - t + 2$

(b) $p(k) = (k-1)(k+1)(k-2)$ — Replacing t by k.

$p(k) = 0 \Rightarrow (k-1)(k+1)(k-2) = 0$ — Do not multiply out the brackets.

$\Rightarrow (k-1) = 0 \text{ or } (k+1) = 0 \text{ or } (k-2) = 0$

$\Rightarrow p(k) = 0 \text{ for } k = 1, -1 \text{ and } 2$

Worked example 3

Two polynomials are given by $P(x) = x^3 + 2x - 1$ and $Q(x) = x^3 - x + 2$.

(a) Simplify $P(x) + 3Q(x)$.

(b) The coefficient of x in the polynomial $R(x)$, given by $R(x) = 2P(x) + kQ(x)$, is zero.

  (i) Find the value of the constant k.

  (ii) Show that for the polynomial $R(x)$, the coefficient of x^3 is equal to the constant term.

(a) $P(x) + 3Q(x) = x^3 + 2x - 1 + 3(x^3 - x + 2)$

$= x^3 + 2x - 1 + 3x^3 - 3x + 6$

$= 4x^3 - x + 5$ — Using **3**

(b) $R(x) = 2P(x) + kQ(x)$

$= 2(x^3 + 2x - 1) + k(x^3 - x + 2)$

$= 2x^3 + 4x - 2 + kx^3 - kx + 2k$

$= (2 + k)x^3 + (4 - k)x + 2k - 2$

  (i) The coefficient of x is $(4 - k) = 0 \Rightarrow k = 4$ — Using **5**

  (ii) The coefficient of x^3 is $(2 + k) = 6$ — Since $k = 4$.

  The constant term is $2k - 2 = 6$ — Using **5**

  $\Rightarrow$ The coefficient of x^3 is equal to the constant term.

Worked example 4

Find the polynomial $P(x)$ if $(x - 2)^2 \times P(x) \equiv 5x^3 - 17x^2 + 8x + k$, and state the value of the constant k.

$(x^2 - 4x + 4) \times P(x) \equiv 5x^3 - 17x^2 + 8x + k$ Multiplying $(x - 2)$ by $(x - 2)$.

The polynomial $P(x)$ must be linear so let $P(x) = ax + b$

Quadratic $\times$ linear = cubic.

$\quad (x^2 - 4x + 4) \times (ax + b) \equiv 5x^3 - 17x^2 + 8x + k$

Equate coefficients of x^3: $a = 5$

Equate coefficients of x: $-4b + 4a = 8$

$\quad \Rightarrow \quad -4b + 20 = 8$ Since $a = 5$.

$\quad \Rightarrow \quad 4b = 12 \quad \Rightarrow \quad b = 3$

Equate constant terms $4b = k \quad \Rightarrow \quad k = 12$ Since $b = 3$.

$P(x) = 5x + 3$ and $k = 12$.

REVISION EXERCISE 5

5

1 Two of the following expressions are polynomials. Add together the two polynomials.

$$3x^3 + 2x - 4 \qquad 6 - 4x + 2x^2 + x^{-1} \qquad 2 + 3x + x^2 - x^3$$

2 Simplify the following
  - **(a)** $(x^3 + 2x^2 - 7x) + (x^2 - 4) + (3x + 2)$
  - **(b)** $4x^5 - 6x^4 + 7x^2 - 3 + 2(3x + 2)$
  - **(c)** $3x^3 - 5x^2 + 7x - 3(x^2 - 2x + 1)$
  - **(d)** $4y^3 - y^4 + 7y^5 - 2(3y^5 + 4y^4 - 2y^3 + y)$

3 Multiply the two polynomials $x^2 - 2x - 5$ and $3x^2 - 2x + 4$ and simplify your answer.

4 The polynomial $p(x)$ is obtained by multiplying the two polynomials $3x^2 - 1$ and $x^3 + 2x - 4$.
  - **(a)** Find $p(x)$ in a simplified form.
  - **(b)** State the degree of the polynomial $p(x)$.

5 Multiply out the brackets $x(x - 1)(x + 1)$, giving your answer in its simplest form.

6 Multiply out the brackets $(y - 3)(y + 3)(y + 2)$, giving your answer in its simplest form.

7 Multiply out the brackets $(n - 1)(n - 3)(n - 2)$, giving your answer in its simplest form.

8 Find the coefficient of p in the polynomial given by $(3p^2 + 2p - 4) + (7p^2 - 5p)$.

9 Find the coefficient of x^2 in the polynomial given by $(3x^2 - 7x + 4)(2x^2 - 3x + 1)$.

10 For the polynomial given by $(x + 2)(x^2 - 4x + 6)$, show that the coefficient of x is equal to the coefficient of x^2.

11 The coefficient of y^2 in the polynomial given by $(3 - 4y + 2y^2)(3y^2 + 3y + k)$ is 9. Find the value of the constant k.

12 The coefficient of n in the polynomial given by $(7 - 4n + n^2)(3 + kn + 2n^2)$ is 16.

  (a) Find the value of the constant k.

  (b) Find the coefficient of n^3.

  (c) Find the coefficient of n^2.

13 The cubic polynomial p(x) is given by p$(x) = (x - 3)(x^2 + 2x + 4)$. Show that p$(x)$ can be written in the form $x^3 + ax^2 + bx - 12$, where a and b are constants whose values are to be found.

14 Two polynomials are added together and the answer is $3x^3 + 4x^2 - 5x + 6$. One of the polynomials is $2x^3 + x^2 + 7x - 8$. Find the other polynomial.

15 Two polynomials are multiplied together and the answer is $2x^3 + x^2 - 2x - 6$. One of the polynomials is $2x - 3$. Find the other polynomial.

16 Find the polynomial q(x) if
q$(x) \times (x^2 - 4x + 2) \equiv x^3 - 8x^2 + 18x - 8$.

17 Find the polynomial p(x) if
$(3x - 1) \times$ p$(x) \equiv 3x^3 + 5x^2 - 14x + 4$.

18 Find the polynomial R(x) if
$(x - 1)^2 \times$ R$(x) \equiv 7x^3 - x^2 - 19x + k$, and state the value of the constant k.

19 The cubic polynomial p(x) is given by
p$(x) = (x - 1)(2x + 3)(cx + d)$, where c and d are constants.

  (a) Given that the coefficient of x^3 is 2, write down the value of c.

  (b) Given also that the coefficient of x is -1, find the value of d.

  (c) Find the three possible values of k for which p$(k) = 0$.

  (d) Find the coefficient of x^2 in p(x).

20 The polynomial p(x) is given by
p$(x) = (x - 2)(x - 3)(ax + b)$, where a and b are constants. Given that the coefficient of x^2 is -3 and the coefficient of x is -4:

  (a) find the value of a and the value of b,

  (b) find the possible values of k for which p$(k) = 0$.

Test yourself	**What to review**
	If your answer is incorrect:
1 Three polynomials are given by $P(x) = x + 3$, $Q(x) = 2x - 7$ and $R(x) = 5 - x$. Find, as polynomials in descending powers of x: **(a)** $P(x) + Q(x) - R(x)$ **(b)** $P(x) \times Q(x) + R(x)$ **(c)** $P(x) \times Q(x) + Q(x) \times R(x)$.	See pp 27–28 Examples 1 & 3 or review Advancing Maths for AQA C1C2 pp 71–75
2 Expand the brackets and give your answers as polynomials in descending powers of x. State the degree of the polynomials in your answers. **(a)** $(x^4 + 2x^2 - 1)(x + 3)$ **(b)** $(x^3 + 2x^2 - 3x + 1)(2 + 4x - x^2 + x^3)$ **(c)** $(2x - 1)(2x + 1)(3x + 2)$	See p 27 Example 1 or review Advancing Maths for AQA C1C2 pp 74–75
3 Show that the coefficient of x^3 is twice the coefficient of x in the product $(x^3 + 2x^2 + 3x - 1)(2x^2 + x + 2)$.	See p 27 Example 1 or review Advancing Maths for AQA C1C2 p 75
4 Find the polynomial $P(x)$ if $(x - 1) \times P(x) \equiv 2x^3 - 3x^2 - 2x + 3$.	See pp 28–29 Example 4 or review Advancing Maths for AQA C1C2 p 77

5

Test yourself ANSWERS

4 $P(x) = 2x^2 - x - 3$

(c) $12x^3 + 8x^2 - 3x - 2$, degree 3

(b) $x^6 + x^5 - x^4 + 14x^3 - 9x^2 - 2x + 2$, degree 6

2 (a) $x^5 + 3x^4 + 2x^3 + 6x^2 - x + 3$, degree 5

1 (a) $4x - 9$ **(b)** $2x^2 - 2x - 16$ **(c)** $16x - 56$

Factors, remainders and cubic graphs

Key points to remember

1 The **factor theorem** states that $(x - a)$ is a factor of the polynomial $P(x) \Leftrightarrow P(a) = 0$.

2 In factorising a cubic polynomial, once a linear factor has been found, the remaining quadratic factor can be found by equating coefficients.

3 To sketch the graph of a cubic function:

Step 1: Find the sign of the coefficient of x^3. This gives the shape of the graph at the extremities.

Step 2: Find the point where the graph crosses the y-axis by finding the value of y when $x = 0$.

Step 3: Find the point (or points) where the graph crosses the x-axis by finding the value of x when $y = 0$. (If there is a repeated root the graph will touch the x-axis.)

Step 4: Calculate values of y for some values of x. This is particularly useful to determine the quadrant in which the graph might turn close to the y-axis.

Step 5: Complete the sketch of the graph by joining the sections. (Your sketch should show the main features of the graph and also, where possible, values where the graph intersects the coordinate axes.)

4 Polynomial = divisor $\times$ quotient + remainder.

5 When a polynomial is divided by a linear expression the remainder will always be a constant and the quotient will always be one degree less than the polynomial.

6 **The remainder theorem:**
If a polynomial $P(x)$ is divided by $(x - a)$, the remainder is $P(a)$.

Worked example 1

(a) Use the factor theorem to show that $(x + 2)$ is a factor of the polynomial $P(x) = x^3 + 3x^2 + 3x + 2$.

(b) Prove that the equation $P(x) = 0$ has only one real root and state its value.

$x + 2$ is a factor of $P(x)$ if $P(-2) = 0$ Using **1**

$P(-2) = (-2)^3 + 3(-2)^2 + 3(-2) + 2$

$\qquad = -8 + 12 - 6 + 2$

$\qquad = 0 \quad \Rightarrow \quad x + 2$ is a factor of $P(x)$

(b) Now $\quad P(x) = (x + 2)(ax^2 + bx + c) \equiv x^3 + 3x^2 + 3x + 2$

$\Rightarrow \quad x(ax^2 + bx + c) + 2(ax^2 + bx + c) \equiv x^3 + 3x^2 + 3x + 2$

$\Rightarrow \quad ax^3 + (b + 2a)x^2 + (c + 2b)x + 2c \equiv x^3 + 3x^2 + 3x + 2$

Equate coefficients of x^3: $\quad a = 1$ ●——————————— | Using **2** |

Equate coefficients of x^2: $\quad b + 2a = 3 \quad \Rightarrow \quad b + 2 = 3 \quad \Rightarrow \quad b = 1$

Equate constant terms: $\quad 2c = 2 \quad \Rightarrow \quad c = 1$

So $P(x) = (x + 2)(x^2 + x + 1)$

$\quad P(x) = 0 \quad \Rightarrow \quad (x + 2)(x^2 + x + 1) = 0$

$\quad \quad \quad \quad \quad \Rightarrow \quad x + 2 = 0$

Or $x^2 + x + 1 = 0$, no real roots since $b^2 - 4ac = -3 < 0$

So $P(x) = 0$ has only one real root $x = -2$.

Worked example 2

The function f is defined for all values of x by $f(x) = x^3 - 2x^2 - 5x + 6$.

(a) Find the remainder when $f(x)$ is divided by $x - 2$.

(b) Show that $f(-2) = 0$ and find the value of $f(3)$.

(c) Express $x^3 - 2x^2 - 5x + 6$ as a product of three linear factors.

(d) Sketch the curve with equation $y = x^3 - 2x^2 - 5x + 6$, indicating the coordinates of the points where the curve crosses the coordinate axes.

6

(a) $f(x) = x^3 - 2x^2 - 5x + 6$

$\quad f(2) = 2^3 - 2(2)^2 - 5(2) + 6$ ●——————————— | Using **6** |

$\quad \quad = 8 - 8 - 10 + 6 = -4$

When $f(x)$ is divided by $x - 2$ the remainder is -4.

(b) $f(-2) = -8 - 2(4) + 10 + 6 = 0$

$\quad f(3) = 27 - 18 - 15 + 6 = 0$

(c) $f(-2) = 0 \quad \Rightarrow \quad x + 2$ is a factor of $f(x)$ ●——————— | Using **1** |

$\quad f(3) = 0 \quad \Rightarrow \quad x - 3$ is a factor of $f(x)$ ●——————— | Using **1** |

$\quad f(x) = (x + 2)(x - 3)(ax + b)$

$\quad (x^2 - x - 6)(ax + b) = x^3 - 2x^2 - 5x + 6$

Equate coefficients of x^3: $\quad a = 1$ ●——————————— | Using **2** |

Equate constant terms: $\quad -6b = 6 \quad \Rightarrow \quad b = -1$

$\quad f(x) = (x + 2)(x - 3)(x - 1)$

(d) $y = x^3 - 2x^2 - 5x + 6$

The coefficient of x^3 is $+1$ so shape of the graph is the same as $y = x^3$ when x is numerically very large.

| Using **3** Step 1. |

When $x = 0$, $y = 0 - 0 - 0 + 6 = 6$.

| Using **3** Step 2. |

Graph crosses y-axis at $(0, 6)$.

$y = (x + 2)(x - 3)(x - 1)$

When $y = 0$, $x + 2 = 0$ or $x - 3 = 0$ or $x - 1 = 0$

| Using **3** Step 3. |

$\Rightarrow \quad x = -2, 1, 3$

Graph crosses x-axis at $(-2, 0)$, $(1, 0)$ and $(3, 0)$.

When $x = -1$, $y = -1 - 2 + 5 + 6 = 8 > 6$

| Using **3** Step 4. |

$\Rightarrow$ graph turns in the second quadrant.

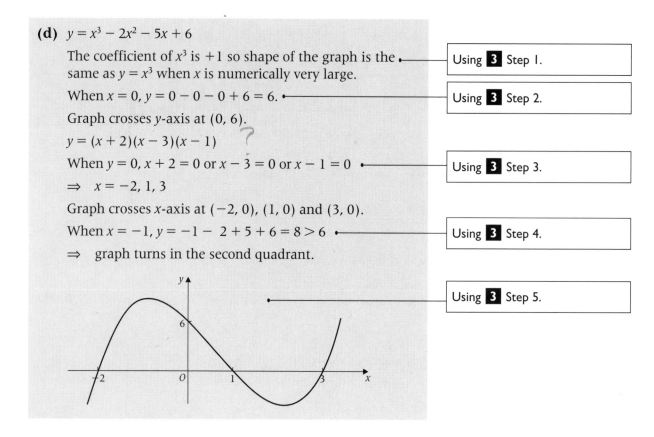

| Using **3** Step 5. |

Worked example 3

The polynomial $P(x) = x^3 + kx^2 + x - 12$, where k is a constant, is such that the remainder on dividing $P(x)$ by $(x + 2)$ is three times the remainder on dividing $P(x)$ by $(x - 2)$.

(a) Find the value of k.

(b) Use the factor theorem to show that $(x - 3)$ is a factor of $P(x)$.

(a) $P(x) = x^3 + kx^2 + x - 12$

When dividing by $x + 2$, remainder $= P(-2)$

| Using **6** |

When dividing by $x - 2$, remainder $= P(2)$

| Using **6** |

$\Rightarrow \qquad P(-2) = 3P(2)$

$-8 + 4k - 2 - 12 = 3[8 + 4k + 2 - 12]$

$\Rightarrow \qquad 4k - 22 = 12k - 6$

$\Rightarrow \qquad -16 = 8k$

$\Rightarrow \qquad k = -2$

(b) $P(x) = x^3 - 2x^2 + x - 12$

$P(3) = 27 - 2 \times 9 + 3 - 12$

| Using **1** |

$\qquad = 30 - 30$

$\qquad = 0 \quad \Rightarrow \quad x - 3$ is a factor of $P(x)$.

Worked example 4

(a) Find the quotient and remainder when $x^3 + 4x + 3$ is divided by $x - 1$.

(b) Hence divide $x^3 + 4x + 3$ by $x - 1$.

(a) $x^3 + 4x + 3 \equiv (x - 1)(x^2 + px + q) + r$, ●────── Using **4** and **5**

where $x^2 + px + q$ is the quotient and r the remainder

$\Rightarrow \quad x^3 + 4x + 3 \equiv x^3 + x^2(p - 1) + x(q - p) + r - q$

Equating coefficients of x^2: $\quad 0 = p - 1 \quad \Rightarrow \quad p = 1$

Equating coefficients of x: $\quad 4 = q - p \quad \Rightarrow \quad 4 = q - 1$
$$\Rightarrow \quad q = 5$$

Equating constant terms: $\quad 3 = r - q \quad \Rightarrow \quad 3 = r - 5$
$$\Rightarrow \quad r = 8$$

The quotient is $x^2 + x + 5$ and the remainder is 8.

(b) $x^3 + 4x + 3 \equiv (x - 1)(x^2 + x + 5) + 8$

Divide both sides by $(x - 1)$ gives

$$\frac{x^3 + 4x + 3}{x - 1} \equiv (x^2 + x + 5) + \frac{8}{x - 1}.$$

6

REVISION EXERCISE 6

1 Use the factor theorem to show that $x - 1$ is a factor of $x^3 + 2x^2 + 4x - 7$.

2 (a) Use the factor theorem to show that $x - 2$ is a factor of $x^3 - 2x^2 + 4x - 8$.

   (b) Hence show that $x^3 - 2x^2 + 4x - 8 = 0$ has only one real root and state its value.

3 (a) The polynomial $P(x) = x^3 - 3x^2 - kx + 8$ has $(x - 4)$ as a factor. Find the value of the constant k.

   (b) Find the value of $P(1)$.

   (c) Write $P(x)$ as a product of three linear factors.

4 The polynomial $f(x) = x^3 + px^2 + qx - 2$ has factors $x - 1$ and $x - 2$.

   (a) Use the factor theorem to find the value of each of the constants p and q.

   (b) Use the remainder theorem to find the remainder when $f(x)$ is divided by $x + 1$.

5 Sketch the graph of the curve with equation $y = (3 - x)(x + 2)^2$.

6 (a) Use the factor theorem to show that $x - 2$ is a factor of $x^3 - 3x^2 + 2x$.

(b) Factorise $x^3 - 3x^2 + 2x$ completely.

(c) Hence find the solutions of the equation $(y + 2)^3 - 3(y + 2)^2 + 2(y + 2) = 0$.

7 (a) Given that $(x + 2)$ is a factor of $P(x) = (x + 5)(x - 2)(x - 1) + k$, find the value of k.

(b) Find the remainder when $P(x)$ is divided by $(x + 1)$.

8 A polynomial is given by $P(y) = (y - 1)(y^2 + sy + t)$. Given that $(y - 4)$ is a factor of $P(y)$ and $(y + 3)$ is a factor of $P(y)$ find the value of each of the constants s and t.

9 A polynomial is given by $P(x) = x^3 + 2x^2 + 3x + 2$.

(a) Use the factor theorem to show that $(x + 1)$ is a factor of $P(x)$.

(b) Express $P(x)$ in the form $(x + 1)Q(x)$, where $Q(x)$ is a quadratic.

(c) By solving the equation $x^3 + 2x^2 + 3x + 2$, show that the curve with equation $y = x^3 + 2x^2 + 3x + 2$ crosses the x-axis at only one point.

10 (a) Use the factor theorem to show that $x - 2$ is a factor of $x^3 - 8x^2 + 17x - 10$.

(b) Factorise $x^3 - 8x^2 + 17x - 10$ completely

(c) Hence solve the equation $x^3 - 8x^2 + 17x - 10 = 0$.

11 (a) Use the factor theorem to show that $x - 4$ is a factor of $x^3 - 3x^2 - 6x + 8$.

(b) Factorise $x^3 - 3x^2 - 6x + 8$ completely.

(c) Hence find the four real roots of the equation $y^6 - 3y^4 - 6y^2 + 8 = 0$.

12 A polynomial $P(x)$ is given by $x^3 - 6x^2 + kx - 8$.

(a) Given that $(x - 2)$ is a factor of $P(x)$, find the value of k.

(b) Factorise $P(x)$ completely.

(c) Hence solve the equation $y^6 - 6y^4 + ky^2 - 8 = 64$.

13 Sketch the graph of $y = (x + 1)^3$.

14 Sketch the graph of $y = x(x^2 - 1)$.

15 Sketch the graph of $y = (2x + 5)(x - 2)^2$.

16 Divide: **(a)** $x + 8$ by x,

(b) $x^2 + 2x + 4$ by $x + 1$,

(c) $x^3 - 2x^2 + 3x - 1$ by $x - 1$.

17 When $x^3 - 2x^2 + 7x + k$ is divided by $x + 1$ the remainder is -8. Find the value of k.

18 The polynomial $P(x) = x^3 - ax^2 + bx + 3$ leaves a remainder of 4 when divided by $x + 1$ and leaves a remainder of 1 when divided by $x - 2$.

   (a) Find the value of a and the value of b.

   (b) Show that $x - 1$ is a factor of $P(x)$.

   (c) Solve the equation $P(x) = 0$, giving your answers in exact form.

19 The polynomial $P(x) = x^3 + kx^2 + x - 4$ leaves a remainder of 2 when divided by $x - 2$. Find the remainder when $P(x)$ is divided by $x - 1$.

20 The polynomial $P(x)$ is defined by $P(x) = x^3 + 3x^2 - 10x - 19$.

   (a) Find the remainder, R, when $P(x)$ is divided by $x + 3$.

   (b) Find the quotient $Q(x)$ such that $P(x) \equiv (x + 3)Q(x) + R$.

6

Test yourself	What to review
	If your answer is incorrect:
1 (a) Use the factor theorem to show that $(x - 2)$ is a factor of $x^3 + 2x^2 - 7x + 2$. **(b)** The polynomial $P(x) = x^3 - 4x^2 + 6x + k$ also has a factor $(x - 2)$. Find the value of the constant k.	See p 32 Example 1 or review Advancing Maths for AQA C1C2 pp 81–82
2 (a) Use the factor theorem to show that $(x - 1)$ is a factor of $P(x) = x^3 + 4x^2 + x - 6$. **(b)** Find the value of $P(-2)$. **(c)** Factorise $x^3 + 4x^2 + x - 6$ completely. **(d)** Hence solve the equation $(y + 1)^3 + 4(y + 1)^2 + (y + 1) - 6 = 0$.	See p 33 Example 2 or review Advancing Maths for AQA C1C2 pp 84–87
3 Sketch the graph of $y = (1 + x)^2(2 - x)$.	See pp 33–34 Example 2 or review Advancing Maths for AQA C1C2 pp 88–90
4 (a) Divide $x + 3$ by $x + 1$. **(b)** Divide by $x^3 + 2x^2 + 4$ by $x + 1$.	See p 35 Example 4 or review Advancing Maths for AQA C1C2 pp 91–93
5 A cubic polynomial $P(x)$ is given by $P(x) = (x - 2)(x - 5)(x - 3)$. Use the remainder theorem to find the remainder when $P(x)$ is divided by **(a)** x **(b)** $x - 4$.	See p 35 Example 4 or review Advancing Maths for AQA C1C2 pp 94–95

1 (b) -4

2 (b) 0 **(c)** $(x-1)(x+2)(x+3)$ **(d)** $y = 0, -3, -4$

3

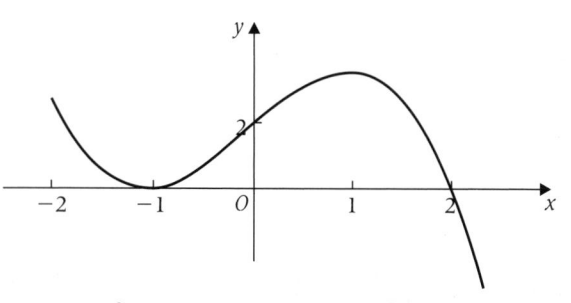

4 (a) $1 + \dfrac{2}{x+1}$ **(b)** $x^2 + x - 1 + \dfrac{5}{x+1}$

5 (a) -30 **(b)** -2

Simultaneous equations and quadratic inequalities

Key points to remember

1 Simultaneous equations can usually be solved by eliminating one of the variables, such as y, and solving for the other variable.

2 When a line intersects a curve, the corresponding equations can be solved simultaneously in order to find the point of intersection.

3 When solving for points of intersection of a straight line and a parabola, it is necessary to solve a quadratic equation of the form $ax^2 + bx + c = 0$.
If $b^2 - 4ac = 0$, the line is a tangent to the parabola.
If $b^2 - 4ac > 0$, the line and parabola intersect at two distinct points.
If $b^2 - 4ac < 0$, the line and parabola do not intersect.

4 Quadratic inequalities can be solved by drawing a graph and considering when the parabola is above or below the x-axis.

5 $y^2 > a^2 \implies y > a$ or $y < -a$

6 $y^2 < a^2 \implies -a < y < a$

7 The values of x for which the quadratic is equal to zero are the critical values. You can easily solve a quadratic inequality by means of a sign diagram, indicating when the expression is positive or negative.

Worked example 1

Solve the simultaneous equations: $y = x - 2$
$$xy + 2y^2 + x^2 = 14$$

$$x(x - 2) + 2(x - 2)^2 + x^2 = 14$$ Substituting $y = x - 2$ in second equation.

$$x^2 - 2x + 2(x^2 - 4x + 4) + x^2 = 14$$ Multiplying brackets.

$$x^2 - 2x + 2x^2 - 8x + 8 + x^2 - 14 = 0$$

$$4x^2 - 10x - 6 = 0$$

$$2(2x + 1)(x - 3) = 0$$ Factorising

$$\implies \qquad x = -\frac{1}{2} \text{ or } x = 3.$$

When $x = -\frac{1}{2}$, $y = -\frac{1}{2} - 2 = -2\frac{1}{2}$ Substituting into $y = x - 2$.

When $x = 3$, $y = 3 - 2 = 1$

The solution is $x = -\frac{1}{2}$ $\Big|$ $x = 3$

The solution is $y = -2\frac{1}{2}$ $\Big|$ $y = 1$

Worked example 2

Calculate the points of intersection of the line with equation $y = 2x - 3$ and the curve with equation $x^2 + y^2 = 2$.

$$x^2 + (2x - 3)^2 = 2$$ — Eliminating y.

$$x^2 + 4x^2 - 12x + 9 = 2$$ — Expanding $(2x - 3)(2x - 3)$.

$$5x^2 - 12x + 7 = 0$$

$$(5x - 7)(x - 1) = 0$$ — Factorising

$$\Rightarrow \quad x = \frac{7}{5} \text{ or } x = 1$$

When $x = \frac{7}{5}$, $y = \frac{14}{5} - 3 = -\frac{1}{5}$ $\Rightarrow$ $\left(\frac{7}{5}, \frac{-1}{5}\right)$ is a point of intersection. — Substituting into $y = 2x - 3$.

When $x = 1$, $y = 2 - 3 = -1$ $\Rightarrow$ $(1, -1)$ is a point of intersection.

Worked example 3

Given that the line $y = x - 1$ is a tangent to the parabola $y = x(x + k)$ find the possible values of k.

$$x - 1 = x^2 + kx$$ — Solving the equations simultaneously.

$$x^2 + (k - 1)x + 1 = 0$$

$$\Rightarrow \quad a = 1, b = k - 1, c = 1$$ — Comparing with $ax^2 + bx + c = 0$.

For the line to be a tangent, $b^2 - 4ac = 0$ or $b^2 = 4ac$ — Using **3**

$$\Rightarrow \quad (k - 1)^2 = 4(1)(1)$$

$$\Rightarrow \quad k - 1 = \pm 2$$

$$\Rightarrow \quad k = 3 \text{ or } k = -1$$

Worked example 4

Solve the inequality $(x + 2)(2x + 1) \leq 9$.

$$2x^2 + 5x + 2 \leq 9$$

$$2x^2 + 5x - 7 \leq 0$$

$$(2x + 7)(x - 1) \leq 0$$ — Factorising

Critical values $x = -3.5$ and $x = 1$

	-3.5		1	
$+$ve		$-$ve		$+$ve

— Indicating sign of $(2x + 7)(x - 1)$ using **7**.

$(2x + 7)(x - 1) \leq 0$ $\Rightarrow$ $-3.5 \leq x \leq 1$

The solution of $(x + 2)(2x + 1) \leq 9$ is $-3.5 \leq x \leq 1$. — Solution includes critical values.

Worked example 5

Find the values of k for which the line $y = x - 2$ intersects the curve with equation $y = x^2 + 3kx + (k + 1)$ at two distinct points.

$$x - 2 = x^2 + 3kx + (k + 1)$$

Using **2**

$$x^2 + (3k - 1)x + k + 3 = 0$$

For two distinct roots, $b^2 - 4ac > 0$

Using **3**

$$(3k - 1)^2 - 4(1)(k + 3) > 0$$

$$9k^2 - 6k + 1 - 4k - 12 > 0$$

$$9k^2 - 10k - 11 > 0$$

$$k^2 - \frac{10}{9}k - \frac{11}{9} > 0$$

$$\left(k - \frac{5}{9}\right)^2 - \frac{11}{9} - \frac{25}{81} > 0$$

$$\left(k - \frac{5}{9}\right)^2 > \frac{124}{81}$$

$$\left(k - \frac{5}{9}\right)^2 > \frac{4 \times 31}{81}$$

$$k - \frac{5}{9} > \frac{2\sqrt{31}}{9} \text{ or } k - \frac{5}{9} < -\frac{2\sqrt{31}}{9}$$

Using **5**

$$\Rightarrow \qquad k > \frac{5 + 2\sqrt{31}}{9} \text{ or } k < \frac{5 - 2\sqrt{31}}{9}$$

7

REVISION EXERCISE 7

1 Solve the following sets of simultaneous equations

   (a) $y = x$ (b) $y = x + 3$
   $\quad\;\; x^2 + y = 6$ $y = 2x^2 - 7$

2 The numbers p and q satisfy the simultaneous equations:
   $p + q = 4$ and $q^2 + 2pq + 9 = 0$.

   (a) Show that $q^2 - 8q - 9 = 0$.

   (b) Hence solve the simultaneous equations.

3 Solve the simultaneous equations: $y = 3x + 1$
   $\qquad\qquad\qquad\qquad\qquad\qquad\quad\; x^2 + 2xy = 5$

4 Solve the simultaneous equations: $x^2 - y^2 = 5xy - 1$
   $\qquad\qquad\qquad\qquad\qquad\qquad\qquad\;\; y = x - 4$

5 Solve the simultaneous equations: $3y^2 - x^2 = \dfrac{xy}{2}$
   $\qquad\qquad\qquad\qquad\qquad\qquad\quad\; x - y = 2$

6 Find the coordinates of the points of intersection of the line
   $y = 2x + 3$ and the curve $y = x(6 - x)$.

7 Find the points of intersection of the curve $y = 7 - x^2$ and the line $2x + y = 4$.

8 Find the coordinates of the points of intersection of the curve $y^2 - x^2 = 5$ and the line $y = -2x + 1$.

9 The line $y = 2x - 7$ intersects the curve with equation $x(x - y) = 10$ at the points S and T.
  (a) Find the coordinates of S and T.
  (b) Show that the distance ST is $3\sqrt{5}$.

10 The line $y = 8 - 2x$ intersects the curve $y = 2x^2 - 3x + 7$ at the points A and B. M is the midpoint of AB
  (a) Find the coordinates of the points A and B.
  (b) Find the coordinates of M.
  (c) Find the equation of the line OM, where O is the origin.

11 Find the value of k such that $y = 2x + k$ is a tangent to the curve with equation $y = x^2 + 1$.

12 Find the values of k such that $y = kx - 2$ is a tangent to the curve $y = x^2 + 1$.

13 (a) Find the values of k such that $y = kx + 1$ is a tangent to the curve with equation $y = x(2 - x)$.
  (b) For each value of k, find the coordinates of the point where the tangent touches the curve.

14 Solve the following inequalities:
  (a) $(x - 1)(x + 2) > 0$ **(b)** $(x - 1)(x + 1) \leqslant 0$
  (c) $x^2 + 2x < 0$ **(d)** $x^2 > 4$

15 Solve the inequality $4x^2 + 5x - 6 < 0$.

16 Solve the inequality $x(x - 2) > 3$.

17 Solve the inequality $x^2 - 2x - 7 \leqslant 0$.

18 Find the set of values of k such that the line $y = kx + (k + 1)$ cuts the curve $y = x^2 - 2x + 3$ in two distinct points.

19 Find the condition of k for the equation $(x + 1)(x^2 + kx + 4) = 0$ to only have one real root.

20 A line has equation $y = mx + 1$, where m is a constant. A curve has equation $y = x^2 - 3x + 10$.
  (a) Show that the x-coordinate of any point of intersection of the line and the curve satisfies the equation $x^2 - (m + 3)x + 9 = 0$.
  (b) Find the values of m for which the equation $x^2 - (m + 3)x + 9 = 0$ has equal roots.
  (c) Describe geometrically the case when m takes either of the values found in part **(b)**.
  (d) Find the set of values of m such that the line $y = mx + 1$ intersects the curve $y = x^2 - 3x + 10$ in two distinct points.

Test yourself	What to review
	If your answer is incorrect:
1 Solve the simultaneous equations $y = 8x + 1$, $2xy = 3$.	See p 39 Example 1 or review Advancing Maths for AQA C1C2 pp 99–101
2 Find the points of intersection of the line $x + y = 5$ and the curve with equation $y = 3x^2 - 25$.	See p 40 Example 2 or review Advancing Maths for AQA C1C2 pp 101–103
3 Prove that the line $y = x$ does not meet the curve with equation $y = (x - 2)(5 - x)$.	See p 40 Example 2 or review Advancing Maths for AQA C1C2 pp 101–103
4 Find the value of k so that the line $y = k - 3x$ is a tangent to the curve with equation $y = 1 + 3x - x^2$.	See p 40 Example 3 or review Advancing Maths for AQA C1C2 p 103
5 Solve the inequality $2x^2 < 3x$.	See p 40 Example 4 or review Advancing Maths for AQA C1C2 pp 105–106
6 Solve the inequality $x^2 + 4x > 1$.	See p 40 Example 4 or review Advancing Maths for AQA C1C2 p 107
7 Find the values of k for which the quadratic equation $4x^2 - 4(k - 1)x + (7k + 1) = 0$ has real roots.	See p 41 Example 5 or review Advancing Maths for AQA C1C2 pp 109–110

7

Test yourself ANSWERS

7 $k \leqslant 0, k \geqslant 9$

6 $x < -2 - \sqrt{5}, x > -2 + \sqrt{5}$

5 $0 > x > 1\frac{1}{2}$

4 $k = 10$

2 $(3, 2), \left(-\dfrac{10}{3}, \dfrac{25}{3}\right)$

1 $x = -\dfrac{1}{2}, y = -3$ and $x = \dfrac{3}{8}, y = 4$

Coordinate geometry of circles

Key points to remember

1   The equation of a circle with centre $(0, 0)$ and radius r is $x^2 + y^2 = r^2$.

2   The equation of a circle with centre (a, b) and radius r is $(x - a)^2 + (y - b)^2 = r^2$.

3   Moving a curve without altering its shape is called a **translation**. The translation vector $\begin{bmatrix} a \\ b \end{bmatrix}$ represents the move a units in the positive x-direction then b units in the positive y-direction.

4   The circle $(x - a)^2 + (y - b)^2 = r^2$ can be obtained from the circle $x^2 + y^2 = r^2$ by applying the translation $\begin{bmatrix} a \\ b \end{bmatrix}$.

5   In general, a translation of $\begin{bmatrix} a \\ b \end{bmatrix}$ transforms the graph of the circle $(x - p)^2 + (y - q)^2 = r^2$ into the graph of $(x - p - a)^2 + (y - q - b)^2 = r^2$.

6   To sketch a circle:
  (i)   find the radius and coordinates of the centre of the circle,
  (ii)   indicate the centre,
  (iii) mark the four points which show the ends of the horizontal and vertical diameters,
  (iv) draw the circle to pass through these four points,
  (v)   if any intercepts with the coordinate axes are integers, normally they should also be indicated.

7   To find the equation of a circle given the coordinates of A and B, the end points of the diameter AB:
  (i)   find the coordinates of the mid-point of AB – this gives the centre $C(a, b)$ of the circle,
  (ii)   find the distance CA (or CB) – this gives the radius r of the circle,
  (iii) use the equation of the circle as $(x - a)^2 + (y - b)^2 = r^2$.

8   The perpendicular bisectors of two chords intersect at the centre of the circle.

9   To determine the conditions for a line to intersect a circle:
  - from the equation of the line make x (or y) the subject,
  - substitute into the equation of the circle to get a quadratic equation in y (or x) of the form $ay^2 + by + c = 0$ (or $ax^2 + bx + c = 0$),
  - **(i)**   if the discriminant $b^2 - 4ac < 0$, there are no real roots and the line does not intersect the circle,
    (ii)   if the discriminant $b^2 - 4ac > 0$, there are two real distinct roots and the line intersects the circle at two points,
    (iii) if the discriminant $b^2 - 4ac = 0$, there is one real (repeated) root and the line touches the circle at one point, the line is a tangent to the circle,
  - if asked to find the coordinates of the points of intersection, solve the quadratic equation and substitute found value(s) into the equation of the line to find the other coordinate(s).

10 The normal at a given point P, on a circle with centre C is the same as the radius CP.

11 To find the equations of tangents and normals to circles at the point $P(x_1, y_1)$:
   (i) find the centre C of the circle,
   (ii) find the gradient, m_1, of CP,
   (iii) the normal is CP so its equation can be found using $y - y_1 = m_1(x - x_1)$,
   (iv) find the gradient, m_2, of the tangent using $m_1 \times m_2 = -1$,
   (v) the equation of the tangent is $y - y_1 = m_2(x - x_1)$.

Worked example 1

A circle with centre C and radius r has equation
$x^2 + y^2 - 6x - 8y = 0$.

(a) By completing the square, express this equation in the
form $(x - a)^2 + (y - b)^2 = k$.

(b) Hence write down:
   (i) the coordinates of C,   **(ii)** the radius of the circle.

(c) The point A, with coordinates $(-1, 7)$, lies on the circle. AB is
a diameter of the circle. Find the coordinates of the point B.

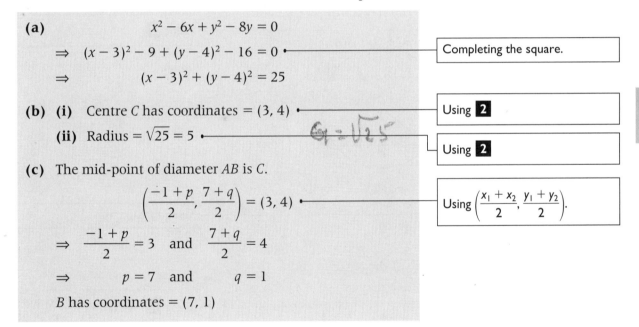

(a) $x^2 - 6x + y^2 - 8y = 0$

   $\Rightarrow (x - 3)^2 - 9 + (y - 4)^2 - 16 = 0$ ⟶ Completing the square.

   $\Rightarrow \qquad (x - 3)^2 + (y - 4)^2 = 25$

(b) (i) Centre C has coordinates $= (3, 4)$ ⟶ Using **2**

   (ii) Radius $= \sqrt{25} = 5$ ⟶ Using **2**

(c) The mid-point of diameter AB is C.

   $\left(\dfrac{-1 + p}{2}, \dfrac{7 + q}{2} \right) = (3, 4)$ ⟶ Using $\left(\dfrac{x_1 + x_2}{2}, \dfrac{y_1 + y_2}{2} \right)$.

   $\Rightarrow \dfrac{-1 + p}{2} = 3$ and $\dfrac{7 + q}{2} = 4$

   $\Rightarrow \qquad p = 7$ and $\qquad q = 1$

   B has coordinates $= (7, 1)$

8

Worked example 2

A circle C has centre $(-1, 3)$ and radius 4.

(a) Write down an equation of C.

(b) The circle C can be obtained from the circle with equation
   $x^2 + y^2 = k$ by applying a translation.
   (i) Find the value of the constant k.
   (ii) Describe the translation.

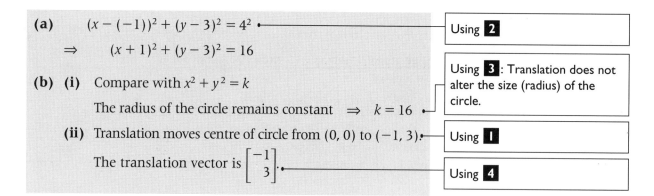

(a) $(x - (-1))^2 + (y - 3)^2 = 4^2$

$\Rightarrow \quad (x + 1)^2 + (y - 3)^2 = 16$

> Using **2**

(b) (i) Compare with $x^2 + y^2 = k$

The radius of the circle remains constant $\Rightarrow k = 16$

> Using **3**: Translation does not alter the size (radius) of the circle.

(ii) Translation moves centre of circle from $(0, 0)$ to $(-1, 3)$.

> Using **1**

The translation vector is $\begin{bmatrix} -1 \\ 3 \end{bmatrix}$.

> Using **4**

Worked example 3

A circle has equation $(x - 1)^2 + (y - 2)^2 = 100$. The point $A(-5, -6)$ lies on the circle and a chord, AB, is of length 12.

(a) Find the distance from the centre of the circle to chord AB.

(b) Show that the equation of the tangent to the circle at A is $3x + 4y + 39 = 0$.

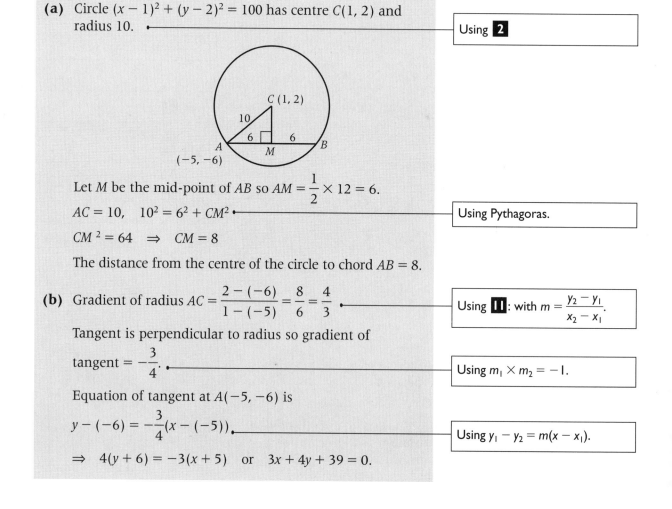

(a) Circle $(x - 1)^2 + (y - 2)^2 = 100$ has centre $C(1, 2)$ and radius 10.

> Using **2**

Let M be the mid-point of AB so $AM = \frac{1}{2} \times 12 = 6$.

$AC = 10, \quad 10^2 = 6^2 + CM^2$

> Using Pythagoras.

$CM^2 = 64 \quad \Rightarrow \quad CM = 8$

The distance from the centre of the circle to chord $AB = 8$.

(b) Gradient of radius $AC = \dfrac{2 - (-6)}{1 - (-5)} = \dfrac{8}{6} = \dfrac{4}{3}$

> Using **11**: with $m = \dfrac{y_2 - y_1}{x_2 - x_1}$.

Tangent is perpendicular to radius so gradient of tangent $= -\dfrac{3}{4}$.

> Using $m_1 \times m_2 = -1$.

Equation of tangent at $A(-5, -6)$ is

$y - (-6) = -\dfrac{3}{4}(x - (-5))$

> Using $y_1 - y_2 = m(x - x_1)$.

$\Rightarrow \quad 4(y + 6) = -3(x + 5) \quad \text{or} \quad 3x + 4y + 39 = 0.$

Worked example 4

A circle C has equation $x^2 + y^2 - 2x - 4y = 0$.

(a) Find the values of m for which the line with equation $y = mx + 9$ does **not** intersect C.

(b) Hence find the equations of the tangents to C from the point $(0, 9)$.

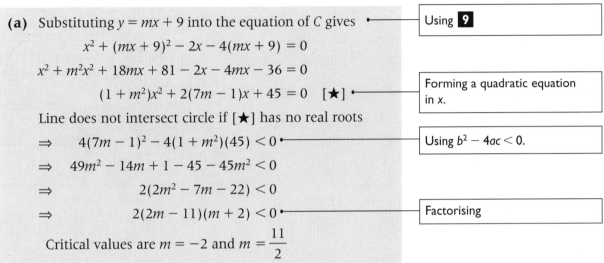

(a) Substituting $y = mx + 9$ into the equation of C gives

$$x^2 + (mx + 9)^2 - 2x - 4(mx + 9) = 0$$

$$x^2 + m^2x^2 + 18mx + 81 - 2x - 4mx - 36 = 0$$

$$(1 + m^2)x^2 + 2(7m - 1)x + 45 = 0 \quad [\bigstar]$$

Line does not intersect circle if $[\bigstar]$ has no real roots

$$\Rightarrow \quad 4(7m - 1)^2 - 4(1 + m^2)(45) < 0$$

$$\Rightarrow \quad 49m^2 - 14m + 1 - 45 - 45m^2 < 0$$

$$\Rightarrow \quad 2(2m^2 - 7m - 22) < 0$$

$$\Rightarrow \quad 2(2m - 11)(m + 2) < 0$$

Critical values are $m = -2$ and $m = \dfrac{11}{2}$

$$(2m - 11)(m + 2)$$

+ve		$-$ve		+ve
	-2		$\frac{11}{2}$	

No real roots for $-2 < m < \dfrac{11}{2}$.

Line does **not** intersect circle if $-2 < m < \dfrac{11}{2}$.

(b) Line touches circle at one point when $b^2 - 4ac = 0$,

$$\Rightarrow \quad m = -2 \text{ and } m = \frac{11}{2}.$$

Equations of tangents from $(0, 9)$ to C are

$$y = -2x + 9 \text{ and } y = \frac{11}{2}x + 9.$$

Using **9**

Forming a quadratic equation in x.

Using $b^2 - 4ac < 0$.

Factorising

8

Worked example 5

A circle has equation $(x - 2)^2 + (y + 1)^2 = 25$ and a straight line L has equation $y = 2x - 5$.

(a) Verify that L passes through the centre, C, of the circle.

(b) The line L intersects the y-axis at the point A. By considering the distance AC, determine whether the point A lies inside the circle.

(c) The point $B(6, -4)$ lies on the circle. Find the equation of the normal to the circle at B, giving your answer in the form $3x + py = q$, where p and q are positive integers.

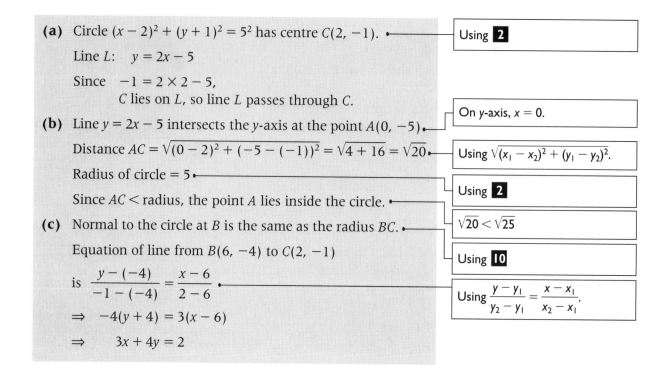

(a) Circle $(x - 2)^2 + (y + 1)^2 = 5^2$ has centre $C(2, -1)$. — Using **2**

Line L: $y = 2x - 5$

Since $-1 = 2 \times 2 - 5$,
        C lies on L, so line L passes through C.

(b) Line $y = 2x - 5$ intersects the y-axis at the point $A(0, -5)$. — On y-axis, $x = 0$.

Distance $AC = \sqrt{(0 - 2)^2 + (-5 - (-1))^2} = \sqrt{4 + 16} = \sqrt{20}$ — Using $\sqrt{(x_1 - x_2)^2 + (y_1 - y_2)^2}$.

Radius of circle $= 5$ — Using **2**

Since $AC <$ radius, the point A lies inside the circle. — $\sqrt{20} < \sqrt{25}$

(c) Normal to the circle at B is the same as the radius BC. — Using **10**

Equation of line from $B(6, -4)$ to $C(2, -1)$

is $\dfrac{y - (-4)}{-1 - (-4)} = \dfrac{x - 6}{2 - 6}$ — Using $\dfrac{y - y_1}{y_2 - y_1} = \dfrac{x - x_1}{x_2 - x_1}$.

$\Rightarrow \quad -4(y + 4) = 3(x - 6)$

$\Rightarrow \qquad 3x + 4y = 2$

REVISION EXERCISE 8

1 (a) Find an equation of the circle with centre $(3, 2)$ and radius 6.
  (b) Write down the equations of the two horizontal tangents to this circle.

2 The point $(1, 0)$ lies on the circle with centre $(-1, 2)$.
  (a) Find the radius of the circle, giving your answer in the form $p\sqrt{2}$.
  (b) Find the equation of the circle in the form $x^2 + y^2 + ax + by + c = 0$, where a, b and c are integers.

3 A circle has equation $x^2 + y^2 - 4x - 4y = 24$.
  (a) By completing the square, express the equation $x^2 + y^2 - 4x - 4y = 24$ in the form $(x - a)^2 + (y - b)^2 = k$.
  (b) (i) State the radius of the circle.
      (ii) Explain why the line with equation $y = x$ is a diameter of the circle.
  (c) Find the coordinates of the endpoints of the diameter whose equation is $y = x$.

4 The area of a circle with centre $(4, -3)$ is 16π.
  (a) Find an equation of the circle.
  (b) Find the equations of the two vertical tangents to the circle.

5 A circle C_1 has equation $x^2 + y^2 + 2x - 8y + 13 = 0$.
  A circle C_2 has equation $(x - 1)^2 + (y - 5)^2 = 4$.
  (a) (i) Find the coordinates of the centre of C_1.
      (ii) Show that C_1 and C_2 have equal radii.
  (b) Describe the geometrical transformation which has been applied to C_1 to obtain C_2.

6 A circle passes through the points $O(0, 0)$, $A(6, 0)$ and $B(4, 4)$.

    (a) Find an equation of the perpendicular bisector of the chord AB.

    (b) Find an equation of the circle.

    (c) Find an equation of the tangent to the circle at B.

7 The lines $y = 2x + 1$ and $y = x + 3$ intersect at the centre of a circle of radius 3.

    (a) Find an equation of the circle.

    (b) Show that the circle does not intersect the x-axis.

8 A circle has equation $(x + 2)^2 + (y + 3)^2 = 49$. A chord, PQ, of the circle is of length 10.

    (a) Write down:
        (i) the coordinates of the centre of the circle,
        (ii) the radius of the circle.

    (b) Find, in the form $p\sqrt{6}$, the distance from the centre of the circle to the chord PQ.

9 **(a)** Show that the line $y = x$ and the circle $x^2 + (y - 1)^2 = 5$ intersect and find the coordinates of the points of intersection.

    (b) Find equations of the normals at each of these points of intersection.

    (c) The point $P(1, 3)$ lies on the circle. Find an equation of the tangent to the circle at P.

10 The line $y = x + c$ is a tangent to the circle $x^2 + y^2 + 2x - 7 = 0$.

    (a) Find the possible values of c.

    (b) Find the coordinates of the points where the tangents touch the circle.

8

11 Find the lengths of the tangents from the point $P(7, 4)$ to the circle $(x - 2)^2 + (y + 1)^2 = 9$.

12 The lengths of the tangents from the point $P(6, 8)$ to the circle with centre $(3, 2)$ is $2\sqrt{5}$. Find an equation of the circle.

13 A circle with centre C has equation $x^2 + y^2 - 10x + 6y + 13 = 0$.

    (a) Find the coordinates of C.

    (b) Show that the circle does not intersect the y-axis.

    (c) Find the x-coordinates of the points where the circle intersects the x-axis, giving your answers in surd form.

    (d) Determine whether the point $P(1, -1)$ lies inside the circle.

14 A circle C has equation $x^2 + y^2 - 2x - 4y - 20 = 0$.

    (a) Find the values of m for which the line $y = mx + 9$ does **not** intersect the circle C.

    (b) Hence find the equations of the tangents to C which pass through the point $(0, 9)$.

Test yourself	**What to review**
	If your answer is incorrect:
1 A translation of $\begin{bmatrix} -2 \\ 3 \end{bmatrix}$ transforms the graph of the circle C_1 with centre $(0, 0)$ and radius 3 into the graph of the circle C_2. Find an equation of C_2.	See pp 45–46 Example 2 or review Advancing Maths for AQA C1C2 pp 118–120
2 (a) Find the coordinates of the centre and the radius of the circle, C, whose equation is $x^2 + y^2 + 4y = 32$. **(b)** Find the coordinates of the endpoints of the horizontal diameter of C.	See p 45 Example 1 or review Advancing Maths for AQA C1C2 pp 115–116
3 The vertices of triangle ABC are $A(1, 3)$, $B(2, 4)$ and $C(4, 2)$. **(a) (i)** Find the gradient of AB. **(ii)** Show that angle $ABC = 90°$. **(b)** Find an equation of the circle which passes through the three points A, B and C.	Review Advancing Maths for AQA C1C2 pp 121–123
4 (a) Verify that the point $T(0, -5)$ lies on the circle with equation $(x + 1)^2 + (y + 2)^2 = 10$. **(b)** Find an equation of the normal to the circle at T. **(c)** Find an equation of the tangent to the circle at T. **(d)** Determine whether the point $A(-4, 0)$ lies inside the circle.	See pp 47–48 Example 5 or review Advancing Maths for AQA C1C2 pp 125–127
5 A circle C has equation $x^2 + y^2 + 6x + 4y + 8 = 0$. **(a)** Find the values of c for which the line $y = 2x + c$ is a tangent to C. **(b)** Find the area enclosed by C giving your answer as a multiple of π.	See p 47 Example 4 or review Advancing Maths for AQA C1C2 p 127

Test yourself ANSWERS

5 (a) $6, -1$ **(b)** 5π

4 (b) $y + 3x + 5 = 0$ **(c)** $3y - x + 15 = 0$ **(d)** No, since $\sqrt{13} > \sqrt{10}$

3 (a) (i) 1 **(b)** $x^2 + y^2 - 5x - 5y + 10 = 0$

2 (a) $(0, -2)$, $r = 6$ **(b)** $(-6, -2)$ and $(6, -2)$

1 $(x + 2)^2 + (y - 3)^2 = 9$

CHAPTER 9

Introduction to differentiation: gradient of curves

Key points to remember

1 The gradient of a chord $= \dfrac{\delta y}{\delta x}$, where δx is the 'increase in x' and δy is the 'increase in y'.

2 The tangent to a curve at any point P is the straight line which just touches the curve at the point P.

3 The gradient of a curve $y = f(x)$ at a point P is defined to be equal to the gradient of the tangent to the curve at P.

4 The process of finding the gradient of the tangent to a curve or the derived function is called differentiation.

5 The derivative is commonly denoted by $\dfrac{dy}{dx}$ or $f'(x)$.

$\dfrac{dy}{dx}$ is called the derivative of y with respect to x.

It is the rate of change of y with respect to x.

6 When differentiating, you can use the following strategy:

   (i) if $y = x^n$, then $\dfrac{dy}{dx} = nx^{n-1}$,

   (ii) if $y = cx^n$ (c is a constant), then $\dfrac{dy}{dx} = cnx^{n-1}$,

   (iii) if $y = f(x) \pm g(x)$, then $\dfrac{dy}{dx} = f'(x) \pm g'(x)$.

7 To find the gradient of a curve at a specific point, substitute the x-coordinate of the point into the expression for the derivative.

9

Worked example 1

The point P with coordinates $(3, 6)$ lies on the curve with equation
$y = x^2 - 2x + 3$.

(a) Find the gradient of the chord AB joining the points on the curve with x-coordinates 2 and 4.

(b) Find $\dfrac{dy}{dx}$.

(c) Show that the tangent at P is parallel to the chord AB.

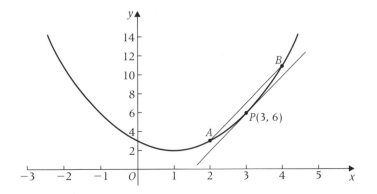

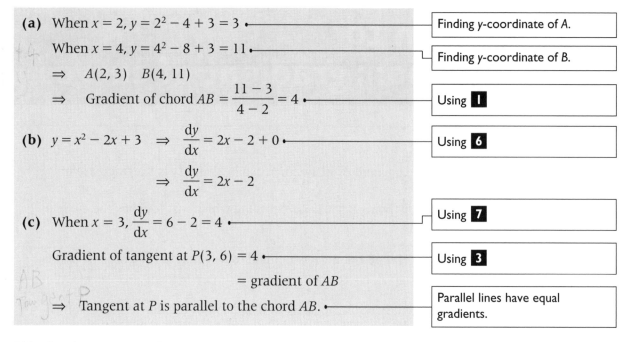

(a) When $x = 2$, $y = 2^2 - 4 + 3 = 3$ — Finding y-coordinate of A.

When $x = 4$, $y = 4^2 - 8 + 3 = 11$ — Finding y-coordinate of B.

$\Rightarrow$ $A(2, 3)$ $B(4, 11)$

$\Rightarrow$ Gradient of chord $AB = \dfrac{11 - 3}{4 - 2} = 4$ — Using **1**

(b) $y = x^2 - 2x + 3$ $\Rightarrow$ $\dfrac{dy}{dx} = 2x - 2 + 0$ — Using **6**

$\Rightarrow$ $\dfrac{dy}{dx} = 2x - 2$

(c) When $x = 3$, $\dfrac{dy}{dx} = 6 - 2 = 4$ — Using **7**

Gradient of tangent at $P(3, 6) = 4$ — Using **3**

$= $ gradient of AB

$\Rightarrow$ Tangent at P is parallel to the chord AB. — Parallel lines have equal gradients.

Worked example 2

(a) It is given that $f(x) = \dfrac{2}{3}x^3 - \dfrac{1}{2}x^2 - 5x - 3$. Find $f'(x)$.

(b) It is given that $g(x) = x^2 + 4x + 7$. Find $g'(x)$.

(c) Hence find the values of x for which $f'(x) = g'(x)$.

(a) $f'(x) = \dfrac{2}{3}(3x^2) - \dfrac{1}{2}(2x) - 5 - 0$ — Using **5** and **6**.

$f'(x) = 2x^2 - x - 5$

(b) $g'(x) = 2x + 4 + 0$ — Using **6**

$g'(x) = 2x + 4$

(c) $2x^2 - x - 5 = 2x + 4$

$2x^2 - 3x - 9 = 0$ — Forming a quadratic equation.

$(2x + 3)(x - 3) = 0$ — Factorising

$\Rightarrow$ $2x + 3 = 0$, $x - 3 = 0$

$\Rightarrow$ $x = -\dfrac{3}{2}$, $x = 3$

Worked example 3

The equation of a curve is $y = x^3 + 2x^2 - 5x - 2$.

(a) Find the gradient of the curve at the point $P(2, 4)$.

(b) Find the coordinates of the point Q on the curve at which the gradient of the curve is the same as the gradient at P.

(a) $\dfrac{dy}{dx} = 3x^2 + 4x - 5$ •————————————— Using **6**

At P, $x = 2$ $\Rightarrow$ $\dfrac{dy}{dx} = 12 + 8 - 5 = 15$ •————— Using **7**

The gradient of the curve at $P = 15$

(b) $\dfrac{dy}{dx} = 15$ $\Rightarrow$ $3x^2 + 4x - 5 = 15$

$\Rightarrow$ $3x^2 + 4x - 20 = 0$

$\Rightarrow$ $(3x + 10)(x - 2) = 0$ •————————— Factorising

$\Rightarrow$ $x = -\dfrac{10}{3}, x = 2$ (point P)

At Q, $x = -\dfrac{10}{3}$ $\Rightarrow$ $y = -\dfrac{1000}{27} + 2\left(\dfrac{100}{9}\right) - 5\left(-\dfrac{10}{3}\right) - 2$

$= -\dfrac{4}{27}.$

Q has coordinates $\left(-\dfrac{10}{3}, -\dfrac{4}{27}\right)$.

Worked example 4

Points A and B are two points on the curve with equation
$y = 4 - 7x + 6x^2 - x^3$. The tangent at A and the tangent at B are
both perpendicular to the line with equation $2y + x = 8$.
Find the coordinates of the points A and B given that A is closer
to the origin than B.

$\dfrac{dy}{dx} = -7 + 12x - 3x^2$ •————————————— Differentiating using **6**.

$2y + x = 8$ $\Rightarrow$ $y = -\dfrac{1}{2}x + 4$ •————————— Rearranging into form $y = mx + c$.

$\Rightarrow$ Gradient of line $= -\dfrac{1}{2}$

$\Rightarrow$ Gradients of tangent at both A and $B = 2$ •————— Using $m_1 \times m_2 = -1$.

$\Rightarrow$ $\dfrac{dy}{dx} = 2$ at points A and B.

$\Rightarrow$ $2 = -7 + 12x - 3x^2$

$\Rightarrow$ $3x^2 - 12x + 9 = 0$

$\Rightarrow$ $3(x^2 - 4x + 3) = 0$ $\Rightarrow$ $3(x - 1)(x - 3) = 0$ •————— Factorising

$\Rightarrow$ $x = 1, x = 3$.

When $x = 1$, $y = 4 - 7 + 6 - 1 = 2$

When $x = 3$, $y = 4 - 21 + 54 - 27 = 10$

Since $(1, 2)$ is closer to the origin than $(3, 10)$, A has
coordinates $(1, 2)$ and B has coordinates $(3, 10)$.

9

Worked example 5

The equation of a curve is $y = (x - 1)(2x + 1)(2x - 1)$.
Find the gradient of the curve at each of the points where the curve crosses the x-axis.

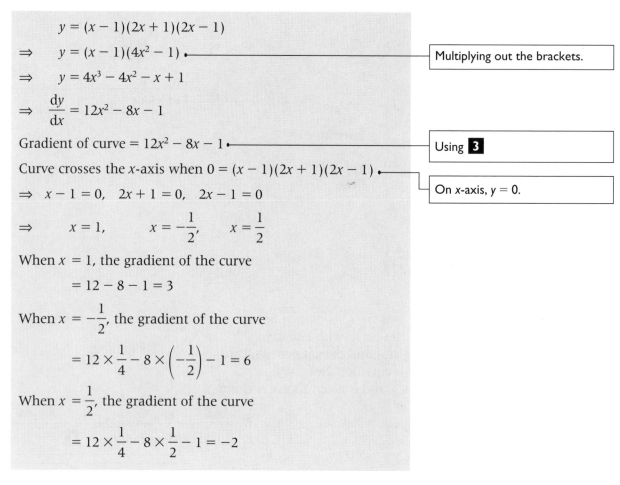

$$y = (x - 1)(2x + 1)(2x - 1)$$

$\Rightarrow \quad y = (x - 1)(4x^2 - 1)$ ———— Multiplying out the brackets.

$\Rightarrow \quad y = 4x^3 - 4x^2 - x + 1$

$\Rightarrow \quad \dfrac{dy}{dx} = 12x^2 - 8x - 1$

Gradient of curve $= 12x^2 - 8x - 1$ ———— Using **3**

Curve crosses the x-axis when $0 = (x - 1)(2x + 1)(2x - 1)$ ———— On x-axis, $y = 0$.

$\Rightarrow \quad x - 1 = 0, \quad 2x + 1 = 0, \quad 2x - 1 = 0$

$\Rightarrow \qquad x = 1, \qquad x = -\dfrac{1}{2}, \qquad x = \dfrac{1}{2}$

When $x = 1$, the gradient of the curve

$\qquad = 12 - 8 - 1 = 3$

When $x = -\dfrac{1}{2}$, the gradient of the curve

$\qquad = 12 \times \dfrac{1}{4} - 8 \times \left(-\dfrac{1}{2}\right) - 1 = 6$

When $x = \dfrac{1}{2}$, the gradient of the curve

$\qquad = 12 \times \dfrac{1}{4} - 8 \times \dfrac{1}{2} - 1 = -2$

REVISION EXERCISE 9

1 A curve has equation $y = x^3 + x$. Find the gradient of the chord joining the points with x-coordinates 1 and 2.

2 Show that the chord joining the points with x-coordinates -1 and 3 on the curve $y = 1 + 4x - x^2$ is parallel to the line $2y - 4x = 7$.

3 The chord joining the points with x-coordinates 0 and 2 on the curve $y = 3 + kx - x^2$ is parallel to the line $y = 2x + 1$. Find the value of the constant k.

4 Find the derivative with respect to t for each of the following:

   (a) $y = 4 + t^2$ **(b)** $x = 3 - 7t + \dfrac{1}{2}t^2$ **(c)** $A = (t + 2)^2$

5 Find $\dfrac{dy}{dx}$ for each of the following:

(a) $y = x(x + 1)$

(b) $y = x + 3 + 2(x + 1)(x - 1)$

(c) $y = (x - 2)(x^2 - x + 1)$

(d) $y = \dfrac{1}{2}(2x + 1)(x + 4)$

(e) $y = \dfrac{3}{5}x^5 - \dfrac{1}{2}x^4 + \dfrac{2}{3}x^3 + (x + 1)^2$

6 A curve has equation $y = 2 - 3x + 2x^2$.

(a) Find the gradient of the curve at the point $P(2, 4)$.

(b) Find the gradient of the curve at the point Q where $x = -1$.

(c) Find the gradient of the chord PQ.

7 The equation of a curve is $y = (x - 2)(x + 3)$.

(a) Find the gradient of the curve:

(i) at the point where the curve crosses the y-axis,

(ii) at the points where the curve crosses the x-axis.

(b) The gradient of the curve is -6 at the point P. Find the coordinates of P.

8 Find the coordinates of the points on the curve $y = x^3 - 9x + 5$ where the gradient is equal to 3.

9 Find the coordinates of the points on the curve $y = \dfrac{1}{3}x^3 - 2x^2 - 3x$ at which the tangents are parallel to the line $y + 6x = 1$.

10 Given that $f(x) = 2x^3 - x^2 + 6$, find:

(a) the value of $f(1)$,

(b) (i) $f'(x)$,

(ii) the value of $f'(1)$.

11 The line $y = x$ intersects the curve $y = x^2 - 4x + 4$ at the points A and B, where A is closer to the origin than B.

(a) Find the coordinates of A and B.

(b) Find the gradient of the curve at each of the points A and B.

9

12 Functions f and g are defined by $f(x) = x^4 - x^2 - 4x + 6$ and $g(x) = x(x - 4) + 6$. Find the three values of x for which $f'(x) = g'(x)$.

13 Points A and B lie on the curve with equation $y = x^2 - 8x + 10$. The tangent at the point $A(5, -5)$ is perpendicular to the tangent at B. Find the x-coordinate of B.

14 A curve has equation $y = x(x - 2)^2$.

   (a) Find $\dfrac{dy}{dx}$.

   (b) Verify that $\dfrac{dy}{dx} = -1$ when $x = 1$.

   (c) Find the x-coordinates of the points on the curve at which the gradient of the tangent is zero.

   (d) Hence explain why the x-axis is a tangent to the curve and write down the coordinates of the point where it touches the curve.

15 The equation of a curve is $y = x^3 - 2x^2 - x$.

   (a) Find the gradient of the curve at the point $P(2, -2)$.

   (b) Point Q lies on the curve. The tangent at Q is parallel to the tangent at P. Find the coordinates of Q.

16 A curve has equation $y = \dfrac{1}{2}(3x - 2)^2 - 2x^3$.

   (a) Find $\dfrac{dy}{dx}$.

   (b) Find the gradient of the curve at the point $P\left(1, -\dfrac{3}{2}\right)$.

   (c) Show that the tangent at P is perpendicular to the line $3y - x = 4$.

   (d) The point Q lies on the curve. Given that the tangent at Q is also perpendicular to the line $3y - x = 4$, find the coordinates of Q.

   (e) By writing $\dfrac{dy}{dx}$ in the form $-p[(x - q)^2 + r]$, where p, q and r are positive constants, show that $\dfrac{dy}{dx}$ is never positive.

Test yourself	What to review

If your answer is incorrect:

1 Show that the gradient of the chord joining the points with x-coordinates -1 and 3 on the curve with equation $y = 4 + 5x - 2x^2$ is equal to the gradient of the line $y = x - 7$.

See p 51 Example 1 or review Advancing Maths for AQA C1C2 p 138

2 For the following, find the derivative of y with respect to x:
(a) $y = 4x^5 - 7x + 3$
(b) $y = \dfrac{1}{2}x^4 - \dfrac{2x^3}{3} + x - 2$

See p 52 Example 2 or review Advancing Maths for AQA C1C2 pp 145–146

3 A curve has equation $y = 3 - 2x + x^2 - 2x^3$.
(a) Find the gradient of the curve at the point $(2, -13)$.
(b) Find the gradient of the tangent at the point where the curve crosses the y-axis.

See p 52–53 Example 3 or review Advancing Maths for AQA C1C2 pp 147–148

4 The equation of a curve is $y = \dfrac{3}{2}x^2 - 4x + 7$.
(a) Find the coordinates of the point P, at which the gradient of the curve $= -10$.
(b) Show that P lies on the line $y = 11 - 5x$.

See p 54 Example 5 or review Advancing Maths for AQA C1C2 p 149

5 Find $f'(x)$ in each of the following:
(a) $f(x) = 4x(x - 1)$.
(b) $f(x) = (2x + 5)^2$.
(c) $f(x) = (x + 6)(2x - 3)(2x + 3)$.

See p 52 Example 2 or review Advancing Maths for AQA C1C2 pp 150–151

Test yourself ANSWERS

9

2 (a) $\dfrac{dy}{dx} = 20x^4 - 7$ (b) $\dfrac{dy}{dx} = 2x^3 - 2x^2 + 1$

3 (a) -22 (b) -2

4 (a) $(-2, 21)$

5 (a) $8x - 4$ (b) $8x + 20$ (c) $12x^2 + 48x - 9$

CHAPTER 10

Applications of differentiation: tangents, normals and rates of change

Key points to remember

1 The equation of the tangent to a curve at the point $P(x_1, y_1)$ is given by $y - y_1 = m(x - x_1)$, where m is the gradient of the curve at P.

2 The normal to a curve at the point P is a straight line that is perpendicular to the tangent at P. When the gradient of the tangent is m, the gradient of the normal is $-\dfrac{1}{m}$.

3 $\dfrac{dP}{dQ}$ represents the rate of change of P with respect to Q. A positive gradient represents a rate of increase whereas a negative gradient represents a rate of decrease.

4 $\dfrac{dy}{dx} > 0 \quad \Rightarrow \quad y$ is increasing with respect to x

$\dfrac{dy}{dx} < 0 \quad \Rightarrow \quad y$ is decreasing with respect to x.

5 When $f'(x) > 0$ in a given interval, the function f is said to be an **increasing function** over this interval.

When $f'(x) < 0$ in a given interval, the function f is said to be a **decreasing function** over this interval.

Worked example 1

A curve has equation $y = x^3 - 3x^2 + x$.

(a) Find an equation of the tangent to the curve at the origin.

(b) The normal to the curve at the point where $x = 1$ intersects this tangent at the point P. Find the coordinates of P.

(a) $\dfrac{dy}{dx} = 3x^2 - 6x + 1$

> Differentiating: $\dfrac{d}{dx}(cx^n) = ncx^{n-1}$.

At origin $(0, 0)$, $\dfrac{dy}{dx} = 0 - 0 + 1 = 1$

> Finding the gradient of the tangent at the origin.

Equation of the tangent at the origin is $y - 0 = 1(x - 0)$

$$\text{or} \quad y = x$$

> Using **1**

(b) When $x = 1, y = 1 - 3 + 1 = -1$

> Finding the y-coordinate when the x-coordinate $= 1$.

Gradient of tangent at $(1, -1) = 3 - 6 + 1 = -2$

Gradient of normal at $(1, -1) = -\left(\dfrac{1}{-2}\right) = \dfrac{1}{2}$

> Value of $\dfrac{dy}{dx}$ gives gradient of tangent.

Equation of normal at $(1, -1) = y - (-1) = \dfrac{1}{2}(x - 1)$

> Using **2**

$$\text{or} \qquad 2y = x - 3$$

Normal $2y = x - 3$ intersects tangent $y = x$

> Using $y - y_1 = m(x - x_1)$.

$\Rightarrow \qquad 2x = x - 3$

$\Rightarrow \qquad x = -3$

> Solving simultaneous equations.

When $\quad x = -3, y = -3$

$\Rightarrow \quad$ Point of intersection P has coordinates $(-3, -3)$.

> Substituting in $y = x$.

Worked example 2

The value of an antique is £V after t years, where $V = 400 + (2t + 5)^2$.

(a) Find the value of t when $V = 500$.

(b) Find $\dfrac{dV}{dt}$.

(c) Find the rate of increase of the value of the antique after 10 years.

(a) $\qquad 500 = 400 + (2t + 5)^2$

> Replacing V by 500 in $V = 400 + (2t + 5)^2$.

$\Rightarrow \qquad 100 = (2t + 5)^2$

$\Rightarrow \quad 2t + 5 = \pm 10$

> Taking square root of both sides.

$\Rightarrow \qquad 2t = 10 - 5$

> t cannot be negative.

$\Rightarrow \qquad t = 2.5$

(b) $\qquad V = 400 + 4t^2 + 20t + 25$

> Multiplying out the brackets.

$\qquad V = 4t^2 + 20t + 425$

> Simplifying

$\Rightarrow \quad \dfrac{dV}{dt} = 8t + 20$

> Differentiating V with respect to t.

(c) Need to find the value of $\dfrac{dV}{dt}$ when $t = 10$

> Using **3**

When $t = 10, \quad \dfrac{dV}{dt} = 80 + 20$

$$\dfrac{dV}{dt} = 100$$

Rate of increase after 10 years is £100 per year.

> $\dfrac{dV}{dx} > 0 \Rightarrow$ a rate of **increase**.

10

Worked example 3

The function f is defined for all real values of x by
$f(x) = 10 - 13x + 6x^2 - x^3$.

(a) Find $f'(x)$.

(b) Show that f is a decreasing function.

(a) $f'(x) = -13 + 12x - 3x^2$ —— Differentiating f(x) with respect to x.

(b) $f'(x) = -3(x^2 - 4x) - 13$

$\Rightarrow \quad f'(x) = -3(x^2 - 4x + 4) - 1$ —— Completing the square.

$\Rightarrow \quad f'(x) = -3(x - 2)^2 - 1$

For all real values of x, $(x - 2)^2 \geqslant 0 \quad \Rightarrow \quad -3(x - 2)^2 \leqslant 0$

$\Rightarrow \quad f'(x) \leqslant -1$

So $f'(x)$ is always $< 0 \quad \Rightarrow \quad$ f is a decreasing function. —— Using **5**

Worked example 4

The functions f and g are defined for all real values of x by
$$f(x) = \frac{1}{3}x^3 - x^2 - 15x + 1 \text{ and } g(x) = x^2 - 18x + 11.$$

(a) Find the values of x for which the function f is increasing.

(b) Find the values of: **(i)** $f'(1)$ **(ii)** $g'(1)$.

(c) Find the two values of x at which the functions f and g are decreasing at the same rate.

(a) $f'(x) = \frac{1}{3}(3x^2) - 2x - 15 + 0$ —— Differentiating

$f'(x) = x^2 - 2x - 15$

$f'(x) = (x + 3)(x - 5)$ —— Factorising

f is increasing when $(x + 3)(x - 5) > 0$ —— Using **5**

The critical values are $x = -3$ and $x = 5$ —— Solving $f'(x) = 0$.

$$(x + 3)(x - 5)$$

+ve		−ve		+ve
	−3		5	

The sign diagram for $f'(x) \Rightarrow$ the function f is increasing
for $x < -3, x > 5$.

(b) **(i)** $\quad f'(x) = x^2 - 2x - 15$

$\Rightarrow \quad f'(1) = 1 - 2 - 15 = -16$

(ii) $\quad g'(x) = 2x - 18$ —— Differentiating g(x).

$\Rightarrow \quad g'(1) = 2 - 18 = -16$

(c) f and g are changing at the same rate $\Rightarrow$ f$'(x)$ = g$'(x)$

$\Rightarrow$ $\quad x^2 - 2x - 15 = 2x - 18$

$\Rightarrow$ $\quad x^2 - 4x + 3 = 0$

$\Rightarrow$ $(x - 1)(x - 3) = 0$ •——————————— | Factorising |

$\Rightarrow$ $\quad\quad x = 1, x = 3.$

When $x = 1$, f$'(1)$ = g$'(1)$ = -16 •—————— | From part **(b)**. |

When $x = 3$, f$'(3)$ = g$'(3)$ = -12

$\Rightarrow$ f and g are **decreasing** at the same rate at
$\quad\quad x = 1$ and $x = 3$. •——————— | Rate of change is negative in both cases so the functions are **decreasing** at the same rate. |

REVISION EXERCISE 10

1 A curve has equation $y = 4 + x^2 - 2x^3$.

    (a) Show that the point $(-1, 7)$ lies on the curve.

    (b) Find the value of $\dfrac{dy}{dx}$ when $x = -1$.

    (c) Find an equation of the tangent to the curve at the point $(-1, 7)$.

2 Find the equation of the tangent to the curve
$y = \dfrac{1}{2}x^4 - \dfrac{5}{2}x^2 + x - 3$ at the point where $x = 2$, giving
your answer in the form $y = ax + b$.

3 **(a)** Find an equation of the normal to the curve
$\quad\quad y = x^6 - 2x^5 + x^3 + 2x - 4$ at the point $(1, -2)$.

    (b) Show that the normal intersects the x-axis at the point
$\quad\quad (-1, 0)$.

4 Find an equation of the normal to the curve
$y = x^3 - 2x^2 + x + 1$:

    (a) at the point $(2, 3)$,   **(b)** at the point where $x = 0$.

5 The curve with equation $y = x^2 - 2x - 3$ intersects the x-axis
at the point $(3, 0)$ and at the point A.

    (a) Find the coordinates of A.

    (b) Find an equation of the normal to the curve:
        (i) at the point $(3, 0)$,   **(ii)** at the point A.

    (c) These two normals intersect at the point P. Find the
coordinates of P.

6 Find the equations of the tangents to the curve
$y = x^3 - \dfrac{3}{2}x^2 - 6x + 1$ which are parallel to the x-axis.

10

7 A curve has equation $y = 2x^2 - 3x - 5$. Find the equation of the tangent to the curve which is parallel to the line $y = 5x - 3$.

8 The curve $y = (x^2 + 3)(x - 2)$ intersects the x-axis at the point A and the y-axis at the point B.

   (a) Find the coordinates of: **(i)** A,   **(ii)** B.

   (b) Find $\dfrac{dy}{dx}$.

   (c) Show that the gradient of the curve at A is 7 and find the gradient of the curve at B.

   (d) Show that the equation of the tangent at A is $y = 7x - 14$.

   (e) Show that the tangent at B passes through the point A.

9 Express these statements symbolically:

   (a) The rate of increase of y with respect to x is 3.

   (b) The rate of decrease of x with respect to t is 4.

   (c) V increases at a rate of $(2t^2 + 6)$ with respect to t.

10 Find the rate of change of P with respect to t, where $P = t^5 + 2t^2 + 3$.

11 A ball is projected vertically so that, after t seconds its height, h metres, is given by $h = 20t - 5t^2$, $0 \leqslant t \leqslant 4$.

   (a) Find the rate at which the height is increasing after 1.5 seconds.

   (b) Find the rate of change of h after 3 seconds and comment on the result

   (c) After T seconds the height is increasing at a rate of 8 m s^{-1}. Find the value of T.

12 It is given that $y = t^3 - 4t^2 + 3t + 1$ and $x = t^2 - 4t + 1$. Find the values of t at which the rate of change of y with respect to t is the same as the rate of change of x with respect to t.

13 Show that the function f defined for all real values of x by $f(x) = 8 - 2x - x^7$ is a decreasing function.

14 Find the values of x for which the function g defined by $g(x) = 7 - 4x + x^2$ is decreasing.

15 The function f is defined for all real values of x by $f(x) = x^3 - 6x^2 + 13x - 1$.

   (a) Find $f'(x)$.

   (b) By writing $f'(x)$ in the form $a(x - b)^2 + c$, where a, b, and c are constants, show that f is an increasing function.

Test yourself	**What to review**
	If your answer is incorrect:
1 The point $P(2, 4)$ lies on the curve $y = x^3 - 3x + 2$.	See p 58 Example 1 or review Advancing Maths for AQA C1C2 pp 153–154
(a) Find the equation of the tangent to the curve at the point P, giving your answer in the form $y = ax + b$.	
(b) The tangent at the point Q on the curve is parallel to the tangent at P. Find the coordinates of Q.	
2 The normal to the curve $y = x^4 - 3x^2 + 4$ at the point where $x = 1$ intersects the line $y = x + 3$ at the point A.	See p 58 Example 1 or review Advancing Maths for AQA C1C2 pp 155–156
(a) Find an equation of the normal.	
(b) Find the coordinates of A.	
3 The volume, $V\,\mathrm{m}^3$, of oil in a container is given by $V = \dfrac{35}{4} - \dfrac{1}{4}t^2 - \dfrac{1}{2}t$, where t is the time in hours measured from noon and $0 \leqslant t \leqslant 5$.	See p 59 Example 2 or review Advancing Maths for AQA C1C2 pp 157–160
(a) Find the volume of oil in the container at noon.	
(b) Find the value of $\dfrac{\mathrm{d}V}{\mathrm{d}t}$ when $t = 2$ and interpret the result.	
(c) Find the time of day at which the volume is **decreasing** at a rate of $2.25\,\mathrm{m}^3\,\mathrm{h}^{-1}$.	
4 The function f is defined for all real values of x by $\mathrm{f}(x) = 2x^5 + \dfrac{1}{3}x^3 - 3$.	See p 60 Example 3 or review Advancing Maths for AQA C1C2 pp 161–163
(a) Find the value of $\mathrm{f}'(1)$.	
(b) Show that f is an increasing function.	

Test yourself ANSWERS

4 (a) 11

(c) 3:30 p.m.

(b) -1.5, V is decreasing at a rate of $1.5\,\mathrm{m}^3\,\mathrm{h}^{-1}$

3 (a) $8.75\,\mathrm{m}^3$

2 (a) $2y = x + 3$ **(b)** $(-3, 0)$

1 (a) $y = 9x - 14$ **(b)** $(-2, 0)$

10

CHAPTER 11

Maximum and minimum points and optimisation problems

Key points to remember

1 Stationary points occur when $\dfrac{dy}{dx} = 0$.

2 There are three types of stationary point:
  (i) maximum point,
  (ii) minimum point,
  (iii) stationary point of inflection.

> No questions will be set in the examination involving stationary points of inflection.

3 The nature of a stationary point can be determined by considering the value of the gradient just to the left and right of the point.

Gradient changes from negative to zero to positive – minimum point

Gradient changes from positive to zero to negative – maximum point

Gradient changes from positive to zero to positive – stationary point of inflection

Gradient changes from negative to zero to negative – stationary point of inflection

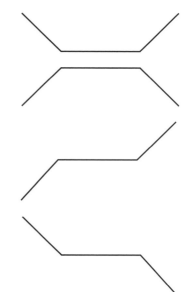

4 The expression obtained by differentiating $\dfrac{dy}{dx}$ is known as the **second derivative** and is denoted by $\dfrac{d^2y}{dx^2}$.

5 If function notation is used then the second derivative of f is denoted by $f''(x)$.

6 If P is a point where $\dfrac{dy}{dx} = 0$ and $\dfrac{d^2y}{dx^2}$ is negative, then P is a maximum point.

7 If Q is a point where $\dfrac{dy}{dx} = 0$ and $\dfrac{d^2y}{dx^2}$ is positive, then Q is a minimum point.

Worked example 1

The curve with equation $y = x^3 + 3x^2 - 9x - 20$ has two stationary points A and B.

Find the equation of the line AB, giving your answer in the form $px + y + q = 0$, where p and q are positive integers.

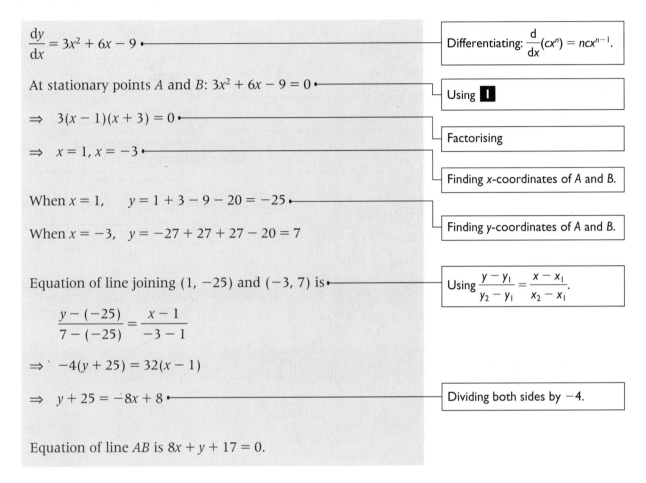

$\dfrac{dy}{dx} = 3x^2 + 6x - 9$

> Differentiating: $\dfrac{d}{dx}(cx^n) = ncx^{n-1}$.

At stationary points A and B: $3x^2 + 6x - 9 = 0$

> Using **1**

$\Rightarrow \quad 3(x - 1)(x + 3) = 0$

> Factorising

$\Rightarrow \quad x = 1, x = -3$

> Finding x-coordinates of A and B.

When $x = 1$, $\quad y = 1 + 3 - 9 - 20 = -25$

When $x = -3$, $\quad y = -27 + 27 + 27 - 20 = 7$

> Finding y-coordinates of A and B.

Equation of line joining $(1, -25)$ and $(-3, 7)$ is

> Using $\dfrac{y - y_1}{y_2 - y_1} = \dfrac{x - x_1}{x_2 - x_1}$.

$$\dfrac{y - (-25)}{7 - (-25)} = \dfrac{x - 1}{-3 - 1}$$

$\Rightarrow \quad -4(y + 25) = 32(x - 1)$

$\Rightarrow \quad y + 25 = -8x + 8$

> Dividing both sides by -4.

Equation of line AB is $8x + y + 17 = 0$.

Worked example 2

The function f is defined for all real values of x by

$$f(x) = \frac{1}{2}(x - 1)^2(1 - 4x).$$

(a) Find $f'(x)$.

(b) **(i)** Find the coordinates of the stationary points of the curve with equation $y = f(x)$.

    (ii) By considering gradients of points close to the stationary points, determine whether they are maximum or minimum points.

(c) Sketch the graph of $y = f(x)$.

11

(a) $f(x) = \dfrac{1}{2}(x^2 - 2x + 1)(1 - 4x)$

$f(x) = \dfrac{1}{2}(-4x^3 + 9x^2 - 6x + 1)$ — Multiplying out the brackets.

$f(x) = -2x^3 + \dfrac{9}{2}x^2 - 3x + \dfrac{1}{2}$

$f'(x) = -6x^2 + 9x - 3$ — Differentiating

(b) (i) At stationary points $-6x^2 + 9x - 3 = 0$ — Using **1**

$\Rightarrow \quad -3(2x - 1)(x - 1) = 0$ — Factorising

$\Rightarrow \quad x = \dfrac{1}{2}, \quad x = 1$ — Finding x-coordinates of stationary points.

When $x = \dfrac{1}{2}, \quad y = \dfrac{1}{2} \times \dfrac{1}{4} \times (-1) = -\dfrac{1}{8}$ — Using $y = \dfrac{1}{2}(x - 1)^2(1 - 4x)$.

When $x = 1, \quad y = \dfrac{1}{2} \times 0 \times (-3) = 0$

Stationary points are $\left(\dfrac{1}{2}, -\dfrac{1}{8}\right)$ and $(1, 0)$.

(ii) Near $x = \dfrac{1}{2}$, — Using $f'(x) = -3(2x - 1)(x - 1)$.

x	0.4	$\dfrac{1}{2}$	0.6
gradient $f'(x)$	-0.36 negative	0 zero	0.24 positive

$\Rightarrow \left(\dfrac{1}{2}, -\dfrac{1}{8}\right)$ is a minimum point. — Using **3**

Near $x = 1$,

x	0.9	1	1.1
gradient $f'(x)$	0.24 positive	0 zero	-0.36 negative

$\Rightarrow (1, 0)$ is a maximum point. — Using **3**

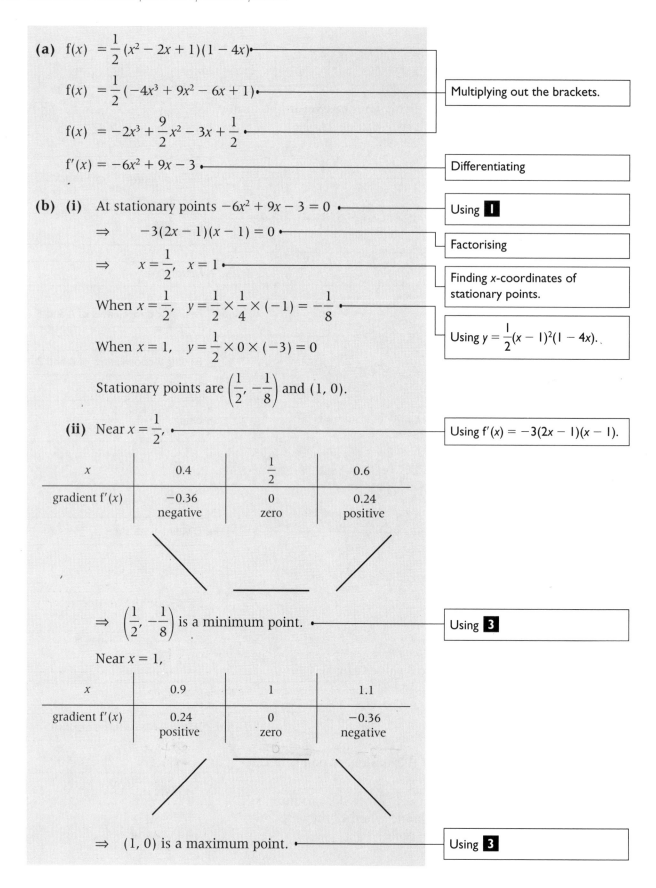

(c) $y = \dfrac{1}{2}(x-1)^2(1-4x)$

When $x = 0$, $y = \dfrac{1}{2}(-1)^2(1) = \dfrac{1}{2}$.

The graph intersects the y-axis at $\left(0, \dfrac{1}{2}\right)$.

When $y = 0$, $\dfrac{1}{2}(x-1)^2(1-4x) = 0$ $\Rightarrow$ $x = 1$ and $x = \dfrac{1}{4}$.

> Find the coordinates of the points where the graph intersects the axes.

The graph intersects the x-axis at $\left(\dfrac{1}{4}, 0\right)$ and $(1, 0)$.

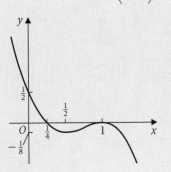

Worked example 3

A curve has equation $y = f(x)$, where $f(x) = (x^2 + 2)^2$.

(a) Show that the curve $y = f(x)$ has only one stationary point, P, and find its coordinates.

(b) Find $f''(x)$.

(c) Use the second derivative test to determine whether the point P is a maximum or a minimum.

(a) $\quad y = f(x) = x^4 + 4x^2 + 4$

> Multiplying out the bracket.

$\quad \dfrac{dy}{dx} = f'(x) = 4x^3 + 8x$

> Differentiating $f(x)$ with respect to x.

$\quad$ For a stationary point $4x^3 + 8x = 0$

> Using **1**

$\quad \Rightarrow \quad 4x(x^2 + 2) = 0$

> Factorising

$\quad \Rightarrow \qquad\qquad x = 0$ only

> Since $x^2 + 2$ is always positive.

$\quad$ When $x = 0$, $y = (0 + 2)^2 = 4$

$\quad$ The only stationary point, P, has coordinates $(0, 4)$.

(b) $\quad f'(x) = 4x^3 + 8x$

$\quad \Rightarrow \quad f''(x) = 12x^2 + 8$

> Using **4** and **5**.

(c) At $P(0, 4)$, $f''(x) = 0 + 8 = 8$.
Since $f''(x)$ is positive at P, the stationary point P is a minimum.

> Using **7**

11

Worked example 4

On a particular flight, a small airline charges a fare of £150 per person plus £5 per person for each unsold seat on its plane. The plane can seat a total of 100 passengers. It is given that x represents the number of unsold seats.

(a) Show that the total revenue, £R, received for the flight is given by $R = 15\,000 + 350x - 5x^2$.

(b) **(i)** Find the value of x for which R has a stationary value.
   (ii) Show that this value of x gives a maximum value.

(c) Hence find the maximum revenue for the flight.

(a) Number of seats sold $= (100 - x)$.
   Fare for each seat $= £(150 + 5x)$.
   Total revenue $= £(100 - x)(150 + 5x)$.
   $\Rightarrow \quad R = 15\,000 + 500x - 150x - 5x^2$
   $\Rightarrow \quad R = 15\,000 + 350x - 5x^2$

(b) **(i)** $\dfrac{\mathrm{d}R}{\mathrm{d}x} = 350 - 10x$ ———— Differentiating

At a stationary point $350 - 10x = 0$ ———— Using **1**
   $\Rightarrow \qquad\qquad\qquad 10x = 350$
   R has a stationary value when $x = 35$.

(ii) $\dfrac{\mathrm{d}^2R}{\mathrm{d}x^2} = -10$ ———— Using **4**

When $x = 35$, $\dfrac{\mathrm{d}^2R}{\mathrm{d}x^2} = -10 < 0 \Rightarrow R$ is a maximum. ———— Using **6**

(c) Maximum value of $R = (100 - 35)(150 + 5 \times 35) = 65 \times 325$ ———— Finding R when $x = 35$.
   Maximum revenue $= £21\,125$.

REVISION EXERCISE 11

1 Find the coordinates of the point on the curve $y = 3x^2 - 12x + 7$ at which the tangent to the curve is horizontal.

2 Find the coordinates of the stationary point on the curve $y = 3 + 2x - 2x^2$.

3 The stationary point on the curve $y = 9 + 10x - 2x^2$ is S.
   The stationary point on the curve $y = (x - 2)(x - 3)$ is T.
   Show that the line ST is vertical.

4 The stationary point on the curve $y = \dfrac{3}{2}x^2 - 6x + 2$ is P.

The stationary point on the curve $y = (x - 1)(x + 3)$ is Q.
   Show that the line PQ is horizontal.

5 The stationary points of the curve $y = x^3 - 2x^2 - 4x - 2$ are A and B. Find the equation of the line AB.

6 Find $\dfrac{d^2y}{dx^2}$ in each of the following:

    (a) $y = x^4 - 3x^2 + 7x$ **(b)** $y = \dfrac{1}{2}x^3 - \dfrac{3}{4}x^2 - 4$

7 For each of the following curves, find the values of $\dfrac{d^2y}{dx^2}$ at the points where the curve crosses the x axis:

    (a) $y = x(x^2 - x - 2)$ **(b)** $y = (x^2 - 1)(x^2 + 2)$

8 For each of the following, find $f''(x)$ when $x = 2$:

    (a) $f(x) = 2x^3 + 3x^2 - 7x + 1$ **(b)** $f(x) = x(x - 1)^2$

9 Find the coordinates of the stationary points of the curve $y = x^2(x - 6)$. Determine the nature of these stationary points.

10 Find the coordinates of the maximum point of the curve $y = 2x^3 + \dfrac{9}{2}x^2 - 6x - 10$.

11 A curve has equation $y = x^3 + \dfrac{3}{2}x^2 - 6x - \dfrac{7}{2}$. The point P is the maximum point of the curve and the point Q is the minimum point.

    (a) Find $\dfrac{d^2y}{dx^2}$.

    (b) Find the coordinates of P and Q.

    (c) The curve with equation $y = (x + a)^2 + b$ has a minimum point at P. Find the values of the integers a and b.

12 The length of a rectangle is x cm. The perimeter of the rectangle is 16 cm and its area is A cm^2.

    (a) Show that $A = 8x - x^2$.

    (b) Show that A has a maximum stationary value.

    (c) Find the maximum area of the rectangle.

    (d) Find, in a simplified surd form, the length of the diagonal of the rectangle when A takes its maximum value.

13 A solid cuboid has length x cm and width $(9 - 2x)$ cm. The height of the cuboid is three times its length and the volume of the cuboid is V cm^3. Given that x varies:

    (a) find an expression for V in terms of x,

    (b) find $\dfrac{d^2V}{dx^2}$,

    (c) **(i)** find the value of x for which V is a maximum,
           (ii) find the maximum possible value of V.

    (d) Find the surface area of the cuboid when its volume takes its maximum value.

11

14 On a touring holiday, a coach firm charges a fare of £200 per person plus £20 per person for each unsold seat on its coach. The coach can carry a total of 50 passengers. Taking x to represent the number of unsold seats:

   (a) Show that the total revenue, £R, received for the touring holiday is given by $R = 10\,000 + 800x - 20x^2$.

   (b) Find the number of unsold seats that will produce the maximum revenue and find the corresponding value of $\dfrac{\mathrm{d}^2R}{\mathrm{d}x^2}$.

   (c) Find the difference in the maximum revenue and the revenue received when all seats on the coach are sold.

Test yourself	**What to review**
	If your answer is incorrect:
1 Show that the curve $y = x^3 + 4x - 5$ has no real stationary points.	See p 67 Example 3 or review Advancing Maths for AQA C1C2 pp 166–169
2 A curve has equation $y = (x^2 + 1)(x - 2)$. Find the value of $\dfrac{\mathrm{d}^2y}{\mathrm{d}x^2}$ at each of the stationary points of the curve.	See p 67 Example 3 or review Advancing Maths for AQA C1C2 pp 171–174

3 The function f is defined for all real values of x by $\mathrm{f}(x) = x^5 - x^3 - 2x + 5$.
  (a) Find $\mathrm{f}''(x)$.
  (b) Solve the equation $\mathrm{f}'(x) = 0$.
  (c) Find the coordinates of the maximum point of the curve $y = \mathrm{f}(x)$.
	See pp 65–66 Example 2 or review Advancing Maths for AQA C1C2 pp 171–174

4 The two shorter sides of a right-angled triangle are x cm and $(20 - 2x)$ cm. The area of the triangle is A cm^2.
  (a) Show that $A = 10x - x^2$.
  (b) Show that A has a maximum stationary value.
  (c) Find the maximum area of the triangle.
  (d) Find, in a simplified surd form, the length of the hypotenuse when A takes its maximum value.
	See p 68 Example 4 or review Advancing Maths for AQA C1C2 pp 177–180

Test yourself ANSWERS

2 $-2, 2$

3 (a) $20x^3 - 6x$ **(b)** $x = \pm 1$ **(c)** $(-1, 7)$

4 (c) 25 cm^2 **(d)** $5\sqrt{5}$ cm

CHAPTER 12

Integration

Key points to remember

1 Integration is the reverse process of differentiation.

2 $\dfrac{dy}{dx} = x^n(n \neq -1) \;\Rightarrow\; y = \dfrac{1}{n+1}x^{n+1} + c$, where c is an arbitrary constant.

3 An indefinite integral has no limits and is of the form $\int f(x)\,dx$. You must remember to include an arbitrary constant in your answer.

4 A definite integral is of the form $\int_a^b f(x)\,dx$, where a and b are the limits.

5 After integrating a definite integral, you obtain $\left[F(x) \right]_a^b$ and this is evaluated as $F(b) - F(a)$.

6 The area under a curve $y = f(x)$ from $x = a$ to $x = b$ is given by $\int_a^b f(x)\,dx$.

7 You can find the area of a region bounded by a curve and a line by finding the areas of separate regions and subtracting one from the other.

8 When a region lies entirely below the x-axis, the value of the definite integral is negative. If the integral has value $-A$, then the area of the region is A.

Worked example 1

Given that $f'(x) = 2x + 3$ and $f(1) = 2$, find the value of $f(2)$.

$f'(x) = 2x + 3$

$\Rightarrow \quad f(x) = 2\dfrac{x^2}{2} + 3x + c$ Using **1** and **2**.

$\quad f(x) = x^2 + 3x + c$

$\quad f(1) = 2 \;\Rightarrow\; 2 = 1 + 3 + c \;\Rightarrow\; c = -2$

$\Rightarrow \quad f(x) = x^2 + 3x - 2$

when $x = 2$, $\quad f(2) = 4 + 6 - 2 = 8$

12

Worked example 2

Find the equation of the curve which has gradient
$\frac{dy}{dx} = 3x^2 - 4x + 3$ and passes through the point of intersection
of the lines $y = x$ and $x + y = 4$.

$y = x$ and $x + y = 4 \Rightarrow x + x = 4$ — Solving equations of lines simultaneously to find the point of intersection.

$\Rightarrow \quad 2x = 4 \Rightarrow x = 2$

$\Rightarrow \quad y = 2$

Curve passes through the point $(2, 2)$

$\frac{dy}{dx} = 3x^2 - 4x + 3$

$\Rightarrow \quad y = 3 \times \frac{x^3}{3} - 4 \times \frac{x^2}{2} + 3x + c$ — Using **2**

$\Rightarrow \quad y = x^3 - 2x^2 + 3x + c$

$2 = 8 - 8 + 6 + c$ — Substituting $x = 2$, $y = 2$ to find c.

$\Rightarrow \quad c = -4$

Equation of curve is $y = x^3 - 2x^2 + 3x - 4$.

Worked example 3

Find $\int 4\left(x + \frac{1}{2}\right)^2 + 15x^2(x^2 + 1) \, dx$.

$\Rightarrow \int 4\left(x^2 + x + \frac{1}{4}\right) + 15x^4 + 15x^2 \, dx$ — Multiplying the brackets.

$\Rightarrow \int 4x^2 + 4x + 1 + 15x^4 + 15x^2 \, dx$

$\Rightarrow \int 15x^4 + 19x^2 + 4x + 1 \, dx$ — Simplifying the expression.

$= \frac{15x^5}{5} + \frac{19x^3}{3} + \frac{4x^2}{2} + x + c$ — Using **2** and **3**.

$= 3x^5 + 6\frac{1}{3}x^3 + 2x^2 + x + c$

Worked example 4

The curve with equation $y = (1 - x)(x - 4)$ intersects the x-axis
at the points P and Q. Calculate the area of the region bounded
by the curve and the x-axis between P and Q.

At P and Q, $y = 0 \Rightarrow 0 = (1-x)(x-4)$

$\Rightarrow x = 1, x = 4$

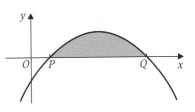

$\text{Area} = \displaystyle\int_1^4 (1-x)(x-4)\,dx$ — Using **6**

$= \displaystyle\int_1^4 -x^2 + 5x - 4\,dx$ — Simplifying the expression.

$= \left[-\dfrac{x^3}{3} + \dfrac{5x^2}{2} - 4x \right]_1^4$ — Using **2**

$= \left(-\dfrac{64}{3} + 40 - 16 \right) - \left(-\dfrac{1}{3} + \dfrac{5}{2} - 4 \right)$ — Using **5**

$= 2\frac{2}{3} - \left(-1\frac{5}{6} \right)$

$= 4\frac{1}{2}$

Worked example 5

The diagram shows a sketch of the curve $y = 9 - x^2$.

The curve intersects the x-axis at points A and B.

The curve intersects the y-axis at the point C.

(a) Find the coordinates of A, B and C.

(b) The region bounded by the curve and the lines AC and BC is shown shaded on the diagram. Find the area of the shaded region.

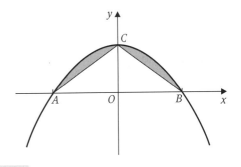

(a) At A and B, $y = 0 \Rightarrow 9 - x^2 = 0 \Rightarrow x^2 = 9 \Rightarrow x = \pm 3$

At C, $x = 0 \Rightarrow y = 9 - 0 = 9$

$A\,(-3,0) \quad B\,(3,0) \quad C\,(0,9)$

(b) Shaded region = area bounded by curve and x-axis from A to B minus area of triangle ACB

$= \displaystyle\int_{-3}^{3} 9 - x^2\,dx - \dfrac{1}{2} \times AB \times CO$ — Using **7**

$= \left[9x - \dfrac{x^3}{3} \right]_{-3}^{3} - \dfrac{1}{2} \times 6 \times 9$

$= \left(27 - \dfrac{27}{3} \right) - \left(-27 + \dfrac{27}{3} \right) - 27$ — Using **5**

$= 18 + 18 - 27 = 9$

12

Worked example 6

The curve $y = 2x^2 - 8x$ intersects the x-axis at the origin and the point T.

(a) Find the coordinates of T and sketch the curve.

(b) The point $S(1, -6)$ lies on the curve. Find the area of the region bounded by the curve and the line ST.

(a) At $T, y = 0 \Rightarrow 2x^2 - 8x = 0$

$\Rightarrow 2x(x - 4) = 0$

$\Rightarrow x = 0, x = 4$

$\Rightarrow T$ has coordinates $(4, 0)$

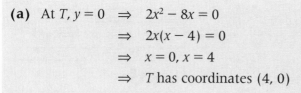

(b)

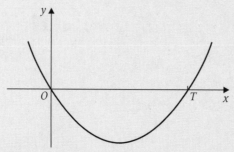

Area of shaded region = area bounded by curve and x-axis from Q to T **minus** area of triangle QST

$$= \int_1^4 2x^2 - 8x \, dx - \frac{1}{2} \times QT \times QS$$

Using **7**

$$= \left[\frac{2x^3}{3} - \frac{8x^2}{2} \right]_1^4 - \frac{1}{2} \times 3 \times 6$$

$$= \left(\frac{128}{3} - 64 \right) - \left(\frac{2}{3} - 4 \right) - 9$$

Using **5**

$$= \frac{126}{3} - 60 - 9$$

$$= -27$$

The negative sign occurs because the region is below the x-axis.
The area of the shaded region is 27.

REVISION EXERCISE 12

1 Find: **(a)** $\int (4x + 1)\, dx,$ **(b)** $\int x^5\, dx,$ **(c)** $\int \frac{x}{2}\, dx.$

2 Given that $\dfrac{dy}{dx} = 2 + 3x$, find y in terms of x.

3 Given that $f'(x) = x^2(x - 4)$, find $f(x)$.

4 Find $\displaystyle\int_1^2 (3x^2 + 2x)\, dx.$

5 Find the area of the finite region bounded by the curve $y = x^3 + 4$, the x-axis and the lines $x = 2$ and $x = 3$.

6 Given that $f'(x) = (x - 1)^2$ and $f(0) = 2$, find $f(x)$.

7 The graph of $y = f(x)$ passes through the point $(-1, 7)$ and $f'(x) = x^3 - 4x^2 + 12x$. Find $f(x)$.

8 Find $\displaystyle\int (x - 2)(4 - 3x)\, dx.$

9 **(a)** Find $\displaystyle\int (x^3 - 12x^2 + 1)\, dx.$

    (b) Hence find the value of $\displaystyle\int_0^1 (x^3 - 12x^2 + 1)\, dx.$

10 **(a)** Find $\displaystyle\int (x - 1)(x + 1)(x + 2)\, dx.$

     (b) Hence find $\displaystyle\int_0^1 (x - 1)(x + 1)(x + 2)\, dx.$

11 Given that $\displaystyle\int_1^k (2x + 1)\, dx = 18$, find the value of the positive constant k.

12 A curve passes through the origin and its gradient at any point (x, y) is given by $\dfrac{dy}{dx} = 8 - \dfrac{3x^2}{2}$.

   (a) Find the equation of the curve.

   (b) Find the coordinates of the points where the curve crosses the x-axis.

   (c) Find the area of the finite region bounded by the curve, the x-axis and the lines $x = 1$ and $x = 2$.

13 Find the area of the finite region bounded by the curve with equation $y = x(2 - x)$, and the line $y = \dfrac{x}{2}$.

14 The line $y = -12$ intersects the curve $y = x(x - 8)$ at the points S and T. Find the area of the finite region bounded by the curve and the line ST.

12

15 The curve with equation $y = x^3 - 9x^2 + 24x + 34$
   is sketched opposite.

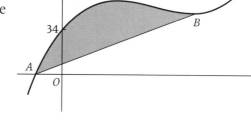

The curve cuts the x-axis at the point A and the
point B is the minimum point of the curve.

   (a) Verify that A has coordinates $(-1, 0)$.

   (b) Show that the x-coordinates of the
   stationary points of the curve satisfy the
   equation $(x - 4)(x - 2) = 0$.

   (c) Find $\dfrac{d^2y}{dx^2}$.

   (d) Find the coordinates of B.

   (e) Find $\displaystyle\int (x^3 - 9x^2 + 24x + 34)\,dx$.

   (f) Find the area of the shaded region bounded by the
   curve and the line AB.

Test yourself	What to review
	If your answer is incorrect:
1 A curve passes through the point $(1, 3)$ and is such that $\dfrac{dy}{dx} = 4x^3 - 6x^2 + 4x + 1$. Find the equation of the curve.	See p 72 Example 2 or review Advancing Maths for AQA C1C2 p 187
2 Find $f(x)$ when $f'(x) = 15x^2(x - 1)(x + 1)$.	See p 71 Example 1 or review Advancing Maths for AQA C1C2 p 188
3 Find $\displaystyle\int_1^2 \dfrac{3x^2}{2} + 2x - 1 \, dx$.	See pp 72–73 Example 4 or review Advancing Maths for AQA C1C2 pp 192–193
4 Find the area of the region bounded by the curve with equation $y = 6x^5 - 2x^3$, the y-axis and the lines $x = 1$ and $x = 2$.	See p72–73 Example 4 or review Advancing Maths for AQA C1C2 pp 194–196
5 Find the area of the finite region bounded by the curve with equation $y = 3x^2 - 2x - 1$ and the x-axis.	See pp 72–73 Example 4 or review Advancing Maths for AQA C1C2 pp 194–196
6 The line $y = x + 7$ intersects the curve $y = (5 - x)(2 + x)$ at the points P and Q. **(a)** Find the coordinates of P and Q. **(b)** Find the area of the finite region bounded by the line PQ and the curve $y = (5 - x)(2 + x)$.	See p 73 Example 5 or review Advancing Maths for AQA C1C2 pp 194–196

1 $y = x^4 - 2x^3 + 2x^2 + x + 1$

2 $f(x) = 3x^5 - 5x^3 + c$

3 5.5

4 55.5

5 $\dfrac{32}{27}$

6 (a) $(-1, 6)$, $(3, 10)$ (b) $10\frac{2}{3}$

Exam style practice paper

Answer **all** questions.

Time allowed: 1 hour 30 minutes

---

1 The point A has coordinates $(4, 3)$ and the point B has coordinates $(8, -3)$.

   (a) Find the gradient of the line AB.   (2 marks)

   (b) Find the coordinates of the mid-point of AB.   (2 marks)

   (c) Hence show that the equation of the perpendicular bisector of AB is $2x - 3y = 12$.   (3 marks)

   (d) This perpendicular bisector intersects the line with equation $3x + 2y = 31$ at the point C. Find the coordinates of C.   (3 marks)

2 (a) Express $(3\sqrt{7} + 5)(2\sqrt{7} - 4)$ in the form $p + q\sqrt{7}$.   (3 marks)

   (b) Simplify $\dfrac{\sqrt{2}}{\sqrt{128} + \sqrt{8}}$.   (3 marks)

3 A curve has equation $y = 3x^2 - 14x + 5$. The point Q lies on the curve and the tangent at Q is parallel to the line with equation $y = x - 42$. Find the x-coordinate of Q.   (4 marks)

4 A circle has equation $x^2 + y^2 + 4x - 6y = 12$.

   (a) (i) Find the coordinates of the centre of the circle.   (3 marks)

       (ii) Find the radius of the circle.   (1 mark)

   (b) Verify that the point $P(-5, 7)$ lies on the circle.   (2 marks)

   (c) Find the gradient of the tangent to the circle at the point P.   (3 marks)

5 The polynomial $p(x)$ is given by $p(x) = x^3 + 3x^2 - 4$.

   (a) Use the remainder theorem to find the remainder when $p(x)$ is divided by x.   (2 marks)

   (b) Use the factor theorem to show that $x + 2$ is a factor of $p(x)$.   (2 marks)

   (c) Factorise $p(x)$ completely and hence solve the equation $p(x) = 0$.   (3 marks)

6 (a) Express $x^2 - 4x + 8$ in the form $(x - p)^2 + q$, where p and q are integers. (2 marks)

(b) A curve has equation $y = x^2 - 4x + 8$. Using your answer to part **(a)**, or otherwise:

  (i) Find the coordinates of the vertex of the curve. (2 marks)

  (ii) State the equation of the line of symmetry of the curve. (1 mark)

(c) Describe geometrically the transformation that maps the graph of $y = x^2 - 4x + 8$ onto the graph of $y = x^2$. (3 marks)

7 The diagram shows the curve C with equation $y = -4x + 4x^2 - x^3$.

The curve touches the x-axis at Q and has a minimum stationary point at P.

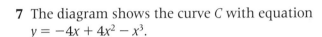

(a) (i) Factorise $-4x + 4x^2 - x^3$. (2 marks)

  (ii) Hence write down the coordinates of Q. (2 marks)

(b) Find $\dfrac{dy}{dx}$. (2 marks)

(c) Find the coordinates of P. (3 marks)

(d) Find the value of $\dfrac{d^2y}{dx^2}$ at P. (2 marks)

(e) (i) Find $\displaystyle\int (-4x + 4x^2 - x^3)\, dx$. (3 marks)

  (ii) Find the area of the shaded region bounded by C and the x-axis. (4 marks)

8 A line has equation $y = k(x - 4) + 1$, where k is a constant. A curve has equation $y = 2x^2 - 6x + 1$.

(a) Show that the x-coordinate of any point of intersection of the line and the curve satisfies the equation $2x^2 - (k + 6)x + 4k = 0$. (1 mark)

(b) The quadratic equation $2x^2 - (k + 6)x + 4k = 0$ has real roots.

  (i) Show that $(k - 2)(k - 18) \geqslant 0$. (5 marks)

  (ii) Hence find the possible values of k. (3 marks)

  (iii) Verify that the line $y = k(x - 4) + 1$ passes through the point $(4, 1)$ for all values of k. (1 mark)

  (iv) Hence find the equations of the tangents from the point $(4, 1)$ to the curve $y = 2x^2 - 6x + 1$, giving your answers in the form $y = mx + c$. (3 marks)

Answers

Revision exercise 1

1 $x = 1\frac{1}{5}$

2 $t = -\frac{1}{6}$

3 $y = 1\frac{2}{5}$

4 $x = 2\frac{1}{4}$

5 $t = -6$

6 $y = 2$

7 $x = 1\frac{1}{2}$

8 $x = -1\frac{2}{5}$

9 $x = 12$

10 $x = 2$

11 $x = 2\frac{2}{5}$

12 $x = -14$

13 $x = 10, y = 2$

14 $x = -11, y = 10$

15 $x = 7, y = 9$

16 $x = -\frac{1}{2}, y = 3$

17 $x = \frac{1}{5}, y = 3\frac{2}{5}$

18 $x = -33, y = -29$

19 $a = -5, b = 1$

20 $a = 2, b = 5$

21 $x < -1$

22 $x > 9$

23 $x \le 1\frac{1}{8}$

24 $x > 6$

25 $x \le 5\frac{1}{2}$

26 $x \le 5\frac{3}{4}$

27 $y < 38$

28 (a) -1

(b) 0

(c) 8

(d) $-\frac{3}{4}$

29 (a) -1

(b) $9\frac{1}{2}$

(c) $-5\frac{1}{2}$

(d) $6\frac{1}{2}$

30 (a) 11

(b) -7

(c) $-3\frac{1}{2}$

(d) $-3\frac{3}{4}$

Revision exercise 2

1 (a) e.g. $\frac{9}{4}, \frac{5}{2}$

(b) e.g. $\sqrt{5}, \sqrt{8}$

2 (a) $2\sqrt{7}$

(b) $3\sqrt{7}$

(c) $4\sqrt{2}$

(d) $5\sqrt{6}$

3 (a) $4 - 2\sqrt{6}$

(b) $3\sqrt{2} - 9$

(c) $6 - 2\sqrt{5}$

(d) $9 + 6\sqrt{2}$

4 (a) $2\sqrt{5}$

(b) $\sqrt{3} - 1$

(c) $\sqrt{5} + 2$

(d) $2 + \sqrt{3}$

5 (a) $8 - 2\sqrt{7}$

(b) $a = \frac{4}{9}, b = \frac{1}{9}$

6 (a) $5 - \sqrt{3}$

(b) $12 + 3\sqrt{3}$

7 $x < -\sqrt{2}$

8 $1 + \sqrt{2}$

9 (a) $\frac{1}{4}(1 - \sqrt{3})$

(b) $k = -2$

10 2

11 $17 - 12\sqrt{2}$

12 $x > 1 + \frac{1}{4}\sqrt{2}$

13 (a) 2

(b) $17 - 12\sqrt{2}$

14 $\frac{1}{2}\sqrt{13}$

15 $\frac{5}{3}$

16 (a) $9(1 + \sqrt{7})$

(b) $x > \frac{1 + \sqrt{7}}{3}$

Revision exercise 3

1 10

2 (a) $M(5, -1)$ **(b)** $\sqrt{26}$

3 (a) -2 **(b)** $\dfrac{1}{2}$ **(c)** Lines are perpendicular

4 (a) $y = 4x + 5$ **(b)** $y = 4x + 8$

5 (a) $y = -2x - 6$ **(b)** $(-2, -2)$

6 (b) $y = -3x$

7 (a) $2\sqrt{13}$ **(c)** $-\dfrac{3}{2}$

8 (a) $k = 14$ **(b)** $p = \dfrac{1}{2}$

9 (a) $\left(\dfrac{3}{2}, -1\right)$ **(b)** $y = -2x + 2$ **(c)** $4y = 2x - 7$

10 (a) $(1, 6)$ **(b)** $2\sqrt{5}$

12 $y = \dfrac{5}{2}x + 4$ **13** $4y = 3x + 15$ **15** $(-13, -10)$ **17** $\dfrac{9}{32}$

18 (a) $\dfrac{7}{10}$ **(b)** $\dfrac{1}{3}$ **(c)** $\dfrac{7}{10}\sqrt{10}$

19 (a) $3y = 4x$ **(b)** $\left(\dfrac{3}{25}, \dfrac{4}{25}\right)$ **(c)** $\dfrac{1}{5}$

Revision exercise 4

1 $(1, 0)$ $(5, 0)$ $(0, 5)$ **2** 6

3 (a) $(0, 0)$ $(1.5, 0)$ **(b)** $x = \dfrac{3}{4}$

4 (a) $(x - 4)^2 - 3$ **(b) (i)** $(4, -3)$ **(ii)** $x = 4$

5 (a) $3(x + 3)^2 - 25$ **(b)** -25 at $x = -3$

6 (a) $(x - 6)(x + 1)$ **(b)** $(0, -6)$ $(6, 0)$ $(-1, 0)$ **(c)** 2.5

7

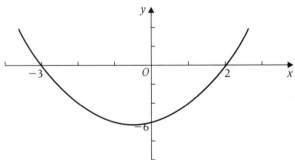

8 4 at $x = 2$

9 (a) $(x + 2)^2 + 1$ **(b)** $(-2, 1)$ **(c)** $x = -2$

10 (a) $\dfrac{2 \pm \sqrt{6}}{2}$ **(b)** $k > -2$

11 $-4, 9$

12 **(a)** translation $\begin{bmatrix} 2 \\ 0 \end{bmatrix}$ **(b)** $(x-2)^2+3$ **(c)** translation $\begin{bmatrix} 2 \\ 3 \end{bmatrix}$

13 **(a)** $y = x^2 + 2x + 1$ **(b)** $y = x^2 + 2$ **(c)** $y = x^2 - 2x + 4$ **(d)** $y = x^2 + 4x - 1$

14 $y = 1 + 2x - x^2$

15 **(a)**

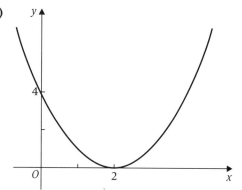

(b) $(x-1)^2$ **(c)** translation $\begin{bmatrix} -1 \\ 0 \end{bmatrix}$

16 **(a)** $(0, -9) \left(-\frac{3}{2}, 0\right) \left(\frac{3}{2}, 0\right)$

(b)

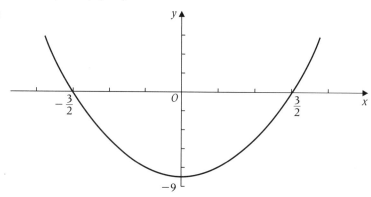

(c) $4x^2 - 8x - 7$

17 **(a)** $(7x + 16)(x - 4)$ **(b)** $\dfrac{-16}{7}, 4$

Revision exercise 5

1 $2x^3 + x^2 + 5x - 2$

2 **(a)** $x^3 + 3x^2 - 4x - 2$ **(b)** $4x^5 - 6x^4 + 7x^2 + 6x + 1$
(c) $3x^3 - 8x^2 + 13x - 3$ **(d)** $y^5 - 9y^4 + 8y^3 - 2y$

3 $3x^4 - 8x^3 - 7x^2 + 2x - 20$

4 **(a)** $3x^5 + 5x^3 - 12x^2 - 2x + 4$ **(b)** 5

5 $x^3 - x$ **6** $y^3 + 2y^2 - 9y - 18$ **7** $n^3 - 6n^2 + 11n - 6$

8 -3 **9** 32 **11** 6

12 **(a)** 4 **(b)** -4 **(c)** 1

13 $a = -1, b = -2$ **14** $x^3 + 3x^2 - 12x + 14$ **15** $x^2 + 2x + 2$

16 $x - 4$ **17** $x^2 + 2x - 4$ **18** $R(x) = 7x + 13, k = 13$

19 **(a)** 1 **(b)** 2 **(c)** $-2, -1.5, 1$ **(d)** 5

20 **(a)** $a = 1, b = 2$ **(b)** $-2, 2, 3$

Revision exercise 6

2 (b) 2

3 (a) 6   **(b)** 0   **(c)** $(x-4)(x-1)(x+2)$

4 (a) $p = -4, q = 5$ **(b)** -12

5

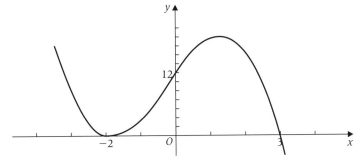

6 (b) $x(x-1)(x-2)$ **(c)** $y = 0, -1, -2$

7 (a) -36 **(b)** -12

8 $s = -1, t = -12$

9 (b) $(x+1)(x^2+x+2)$

10 (b) $(x-2)(x-1)(x-5)$ **(c)** $x = 1, 2, 5$

11 (b) $(x-4)(x-1)(x+2)$ **(c)** $y = \pm 2, \pm 1$

12 (a) $k = 12$ **(b)** $(x-2)^3$ **(c)** $y = \pm\sqrt{6}$

13

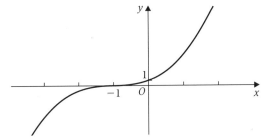

14

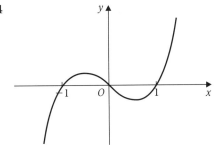

15

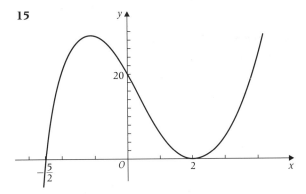

16 (a) $1 + \dfrac{8}{x}$ **(b)** $x + 1 + \dfrac{3}{x+1}$ **(c)** $x^2 - x + 2 + \dfrac{1}{x-1}$

17 2

18 (a) $a = 1, b = -3$ **(b)** $x = -\sqrt{3}, \sqrt{3}, 1$

19 -3

20 (a) 11   **(b)** $x^2 - 10$

Revision exercise 7

1 (a) $x = 2, y = 2$ and $x = -3, y = -3$ (b) $x = -2, y = 1$ and $x = 2.5, y = 5.5$

2 (b) $p = 5, q = -1$ and $p = -5, q = 9$ 3 $x = -1, y = -2$ and $x = \dfrac{5}{7}, y = \dfrac{22}{7}$

4 $x = 5, y = 1$ and $x = \dfrac{3}{5}, y = -\dfrac{17}{5}$ 5 $x = 6, y = 4$ and $x = \dfrac{4}{3}, y = -\dfrac{2}{3}$

6 $(1, 5), (3, 9)$ 7 $(-1, 6), (3, -2)$ 8 $(2, -3), \left(-\dfrac{2}{3}, \dfrac{7}{3}\right)$

9 (a) $(2, -3), (5, 3)$

10 (a) $(1, 6), \left(-\dfrac{1}{2}, 9\right)$ (b) $\left(\dfrac{1}{4}, \dfrac{15}{2}\right)$ (c) $y = 30x$

11 0

12 $\pm 2\sqrt{3}$

13 (a) $0, 4$ (b) $(1, 1), (-1, -3)$

14 (a) $x < -2, x > 1$ (b) $-1 \leqslant x \leqslant 1$

     (c) $-2 < x < 0$ (d) $x < -2, x > 2$

15 $-2 < x < \dfrac{3}{4}$ 16 $x < -1, x > 3$ 17 $1 - 2\sqrt{2} \leqslant x \leqslant 1 + 2\sqrt{2}$

18 $k < -4 - 2\sqrt{5}, k > 2\sqrt{5} - 4$ 19 $-4 < k < 4$

20 (b) $-9, 3$ (c) Line is a tangent to the curve   (d) $m < -9, m > 3$

Revision exercise 8

1 (a) $(x - 3)^2 + (y - 2)^2 = 36$ (b) $y = -4, y = 8$

2 (a) $2\sqrt{2}$ (b) $x^2 + y^2 + 2x - 4y - 3 = 0$

3 (a) $(x - 2)^2 + (y - 2)^2 = 32$

     (b) (i) $4\sqrt{2}$ (ii) centre $(2, 2)$ lies on $y = x$ (since $2 = 2$)

     (c) $(-2, -2), (6, 6)$

4 (a) $(x - 4)^2 + (y + 3)^2 = 16$ (b) $x = 0, x = 8$

5 (a) (i) $(-1, 4)$ (b) translation $\begin{bmatrix} 2 \\ 1 \end{bmatrix}$

6 (a) $2y = x - 1$ (b) $(x - 3)^2 + (y - 1)^2 = 10$ (c) $3y + x = 16$

7 (a) $(x - 2)^2 + (y - 5)^2 = 9$

8 (a) (i) $(-2, -3)$ (ii) 7 (b) $2\sqrt{6}$

9 (a) $(-1, -1)$ and $(2, 2)$ (b) $y = 2x + 1$ and $2y = x + 2$ (c) $2y + x = 7$

10 (a) $-3, 5$ (b) $(1, -2), (-3, 2)$

11 $\sqrt{41}$

12 $(x - 3)^2 + (y - 2)^2 = 25$

13 (a) $(5, -3)$ (c) $5 \pm 2\sqrt{3}$ (d) yes since $\sqrt{20} < \sqrt{21}$

14 (a) $-\dfrac{3}{4} < m < \dfrac{4}{3}$ (b) $y = -\dfrac{3}{4}x + 9, y = \dfrac{4}{3}x + 9$

Revision exercise 9

1 8

3 4

4 (a) $\dfrac{dy}{dt} = 2t$ (b) $\dfrac{dx}{dt} = -7 + t$ (c) $\dfrac{dA}{dt} = 2t + 4$

5 (a) $2x + 1$ (b) $4x + 1$ (c) $3x^2 - 6x + 3$

   (d) $2x + \dfrac{9}{2}$ (e) $3x^4 - 2x^3 + 2x^2 + 2x + 2$

6 (a) 5 (b) -7 (c) -1

7 (a) (i) 1 (ii) -5 and 5 (b) $\left(-\dfrac{7}{2}, \dfrac{11}{4}\right)$

8 $(2, -5)$ and $(-2, 15)$ 9 $\left(1, -4\dfrac{2}{3}\right)$ and $(3, -18)$

10 (a) 7 (b) (i) $6x^2 - 2x$ (ii) 4

11 (a) $A(1, 1), B(4, 4)$ (b) -2 and 4

12 $-1, 0$ and 1 13 $\dfrac{15}{4}$

14 (a) $3x^2 - 8x + 4$ (c) $\dfrac{2}{3}$ and 2 (c) $(2, 0)$

15 (a) 3 (b) $\left(-\dfrac{2}{3}, -\dfrac{14}{27}\right)$

16 (a) $9x - 6 - 6x^2$ (b) -3

   (d) $\left(\dfrac{1}{2}, -\dfrac{1}{8}\right)$ (e) $\dfrac{dy}{dx} = -6\left[\left(x - \dfrac{3}{4}\right)^2 + \dfrac{7}{16}\right]$

Revision exercise 10

1 (b) -8 (b) $8x + y + 1 = 0$

2 $y = 7x - 17$

3 (a) $x + y + 1 = 0$

4 (a) $x + 5y = 17$ (b) $x + y = 1$

5 (a) $(-1, 0)$ (b) (i) $x + 4y = 3$ (ii) $4y = x + 1$ (c) $\left(1, \dfrac{1}{2}\right)$

6 $y = -9, y = 4.5$

7 $y = 5x - 13$

8 (a) (i) $(2, 0)$ (ii) $(0, -6)$ (b) $3x^2 - 4x + 3$ (c) 3

9 (a) $\dfrac{dy}{dx} = 3$ (b) $\dfrac{dx}{dt} = -4$ (c) $\dfrac{dV}{dt} = 2t^2 + 6$

10 $\dfrac{dP}{dt} = 5t^4 + 4t$

11 (a) $5 \, \text{m s}^{-1}$ (b) $-10 \, \text{m s}^{-1}$, height is decreasing at a rate of $10 \, \text{m s}^{-1}$ (c) 1.2

12 $1, \dfrac{7}{3}$

14 $x < 2$

15 (a) $3x^2 - 12x + 13$

Revision exercise 11

1 $(2, -5)$ 2 $(0.5, 3.5)$ 5 $32x + 9y + 26 = 0$

6 (a) $12x^2 - 6$ (b) $3x - \dfrac{3}{2}$

7 (a) -2 at $(0, 0)$, -8 at $(-1, 0)$, 10 at $(2, 0)$ (b) 14 at $(1, 0)$, 14 at $(-1, 0)$

8 (a) 30 (b) 8

9 $(0, 0)$ maximum; $(4, -32)$ minimum

10 $(-2, 4)$

11 (a) $6x + 3$ (b) $P(-2, 6.5)$, $Q(1, -7)$ (c) $a = 2$, $b = 6.5$

12 (c) $16\,\text{cm}^2$ (d) $4\sqrt{2}\,\text{cm}$

13 (a) $27x^2 - 6x^3$ (b) $54 - 36x$

    (c) (i) 3 (ii) 81 (d) $126\,\text{cm}^2$

14 (b) $x = 20$, $\dfrac{\mathrm{d}^2R}{\mathrm{d}x^2} = -40$ (c) £8000

Revision exercise 12

1 (a) $2x^2 + x + c$ (b) $\dfrac{1}{6}x^6 + c$ (c) $y = \dfrac{1}{4}x^2 + c$

2 $y = 2x + \dfrac{3}{2}x^2 + c$

3 $f(x) = \dfrac{1}{4}x^4 - \dfrac{4}{3}x^3 + c$

4 10

5 20.25

6 $f(x) = \dfrac{1}{3}x^3 - x^2 + x + 2$

7 $f(x) = \dfrac{1}{4}x^4 - \dfrac{4}{3}x^3 + 6x^2 - \dfrac{7}{12}$

8 $-x^3 + 5x^2 - 8x + c$

9 (a) $\dfrac{1}{4}x^4 - 4x^3 + x + c$ (b) $-\dfrac{11}{4}$

10 (a) $\dfrac{1}{4}x^4 + \dfrac{2}{3}x^3 - \dfrac{1}{2}x^2 - 2x + c$ (b) $-1\dfrac{7}{12}$

11 $k = 4$

12 (a) $y = 8x - \dfrac{1}{2}x^3$ (b) $(-4, 0)$, $(0, 0)$, $(4, 0)$ (c) $10\dfrac{1}{8}$

13 $\dfrac{9}{16}$

14 $\dfrac{32}{3}$

15 (c) $6x - 18$ (d) $(4, 50)$ (e) $\dfrac{1}{4}x^4 - 3x^3 + 12x^2 + 34x + c$ (f) 93.75

Exam style practice paper

1 (a) $-\dfrac{3}{2}$ (b) $(6, 0)$ (c) $(9, 2)$

2 (a) $22 - 2\sqrt{7}$ (b) $\dfrac{1}{10}$

3 2.5

4 (a) (i) $(-2, 3)$ (ii) 5   (c) $\dfrac{3}{4}$

5 (a) -4 (c) $(x - 1)(x + 2)^2, x = -2, x = 1$

6 (a) $(x - 2)^2 + 4$

(b) (i) $(2, 4)$ (ii) $x = 2$ (c) translation $\begin{bmatrix} -2 \\ -4 \end{bmatrix}$

7 (a) (i) $-x(2 - x)^2$ (ii) $(2, 0)$

(b) $-4 + 8x - 3x^2$ (c) $\left(\dfrac{2}{3}, -\dfrac{32}{27} \right)$ (d) 4

(e) (i) $-2x^2 + \dfrac{4}{3}x^3 - \dfrac{1}{4}x^4 + c$ (ii) $\dfrac{4}{3}$

8 (b) (ii) $k \leqslant 2, k \geqslant 18$ (iv) $y = 2x - 7, y = 18x - 71$